Irish farmers could form no effective professional organisations until they got secure title to their land. By the end of the last century some 400,000 farmers secured that ownership but Irish agriculture had no social structure, no political power, no marketing organisation; it was under-capitalised and relatively unskilled. Each generation that followed helped build a modern structure.

The first generation of owners founded the co-operatives and saw the creation of the Department of Agriculture and political independence.

The second quarter-century was one of bare survival for co-operatives, failure to establish representative farm organisations, and stagnation or regression in production and marketing.

Education made possible the Second Spring of the 1950s. Macra na Feirme, the *Irish Farmers Journal* and the Agricultural Research Institute organised farmers and made them articulate.

In the fourth quarter century farmers asserted themselves, seizing the opportunities offered by the EU, increasing output per man threefold, restructuring their co-operatives into billion-pound multinational enterprises dominating the market both in sale of product and purchase of materials. Their organisations spend nearly half their funds on international representation.

Sean Healy was born in Kilmichael, West Cork in 1926. His grandfather was chairman of the Cork Farmers' Union. He joined the staff of Macra na Feirme in 1948, becoming general secretary in 1951. Having guided the launch, through Macra, of the NFA, he became its first general secretary, retiring in 1985.

Louis Smith was born in Co. Cavan in 1923. He studied Economics and History in UCD and Law in King's Inns. He was organiser of co-operatives with the IAOS from 1950 to 1954, when he was recruited by Macra to help develop policy for the new NFA. He remained as Economist for 10 years. Among other activities he was Economic Consultant to Allied Irish Banks from 1972 to 1989. He was lecturer and then Professor in the Department of Political Economy UCD from 1954 to 1990.

Distribution of Co-operatives 1995

Note: Farm Relief co-ops, not shown, are in every county.
Individual co-ops have branches and may give nationwide service.

Farm Organisations in Ireland

A Century of Progress

LOUIS P.F. SMITH & SEAN HEALY

with a foreword by
TJ Maher

FOUR COURTS PRESS

This book was set in 10.5 on 12.5 point Times New Roman
by Woodcote Typesetters for
FOUR COURTS PRESS LTD
Kill Lane, Blackrock, Co. Dublin, Ireland
and in North America
FOUR COURTS PRESS LTD
c/o ISBS, 5804 N.E. Hassalo Street, Portland, OR 97213.

A catalogue record for this title
is available from the British Library.

ISBN 1-85182-227-1

Printed in Ireland
by Colour Books Ltd, Dublin

Foreword

I welcome this record of the struggle of farmers in the Republic to gain recognition for their place in society and their right to control their own industry. It is a long story with many disappointments. Our pride is that the movement begun by Horace Plunkett and relaunched by Macra has grown to full strength in our generation. Our many loyal comrades who gave their time so generously, sacrificing their own families and farms, deserve commemoration. This book is a reminder to the new generation that the strong voluntary organisations they now enjoy in co-operatives, in Macra, and in the farmers' associations were the fruit of hard work and determination – even risking the prison cell.

Who better to write the history than Sean Healy, who spent his working life in the movement from almost the start of Macra through the birth and adolescence of the NFA and the IFA, and Louis Smith, who worked with the co-operatives and then helped the formation of Macra's economic policy and that of the early National Farmers' Assocation?

When Ireland took her place among the nations as a member of the European Union we were fortunate to find a group which appreciated the contribution that voluntary organisations could make to rural development. Irish farmers have benefited greatly by the support and example of our neighbours in strengthening our organisations in Ireland and opening the world to our trade.

TJ Maher

June 1996

Contents

List of Illustrations

Cartoons are courtesy of the *Irish Farmers Journal* unless otherwise stated.

List of Tables

Abbreviations

BGA	Beet Growers' Association
COGECA	Producer Co-operatives of EU
COPA	Committee of Farm Associations in European Union
DD	Dáil Debates
EC	Executive Council Minutes, NFA\IFA
IAOS/ICOS	Irish Co-operatives Organisation Society
ICMSA	Irish Creamery Milk Suppliers Association
IFAP	International Federation of Agricultural Producers.
IFJ	Irish Farmers' Journal
IP	Irish Press
IT	Irish Times
JSSISI	Journal of the Statistical and Social Inquiry Society of Ireland
LMP	Leinster Milk Producers
NC	National Council Minutes, NFA\IFA
NFA/IFA	Irish Farmers' Association
SD	Senate Debates

Department and Minister refer to Department and Minister of Agriculture.

Value of Money

To avoid confusion due to currency changes, all prices are quoted in new, decimal currency – 100 pence to the pound. The value of money has fallen. Three pence (as defined above; roughly seven old pence) in 1935 could buy as much as a pound in 1994, as in this table:

Number of Pence to Equal £1 in 1994 Value

1922	3 pence	1975	22
1935	3	1980	47
1945	5	1985	77
1950	7	1990	91
1965	9	1994	100

Source: Central Bank

Introduction

Irish agriculture emerged from the tunnel of history after World War 2. For the first century of official statistics Gross Output scarcely increased, while output multiplied in neighbouring countries. In the next generation (from the 1950s) Irish output doubled.

Irish society was without structure when Independence was gained in 1922. Kevin O'Higgins described the situation: 'there was no State and no organised forces ... , no police force was functioning through the country, no system of justice was operating, the whole wheels of administration hung idle, battered out of recognition by the clash of rival jurisdictions.'[1] It was not only the State which had to be built. The smaller organisations which combine to make the structure of a democracy hardly existed. As in other Less Developed Countries, the State had to undertake tasks delegated to trade associations in more settled societies.

The Great Depression of the 1930s, deepened by the Economic War with Britain, the main importer of food, was followed by the isolation of wartime neutrality.

The technology for modern agriculture had been available for generations. However, in 1955 there seemed little cause to hope for improvement. Farms were under-capitalised; livestock breeding was unsatisfactory; marketing was chaotic. Policy decisions affecting agriculture were reached without reference to farm opinion. There was no overall farm organisation to express it.

Previous generations of Irish farmers had concentrated their efforts on survival during the Famine; on the struggle for ownership of the land; on national independence. Now the farmers changed fundamentally. Education made a fundamental change in the farmers themselves. Vocational schools led to the foundation of Macra na Feirme (Young Farmers' Clubs) from 1944. Working at the 'grass-roots' these clubs did for Irish farmers what the Folkhighschools did for Denmark. They developed an educated people, articulate and with the confidence and knowledge to organise their industry.

This history begins with the early and unsuccessful efforts of farmers to organise. The first efforts were scattered in regions or limited to specific commodities. Fragmented associations were weak, without united farmer support. The 'graduates' from Macra formed the National Farmers' Association (now IFA) in 1955. Then came the struggle for recognition by Government; by other citizens; by international organisations. Reorganisation of research, marketing and service co-operatives, co-operation with Government followed. Now, in the European Union, 40 per cent of IFA central expenditure is absorbed by foreign relations.

In forty years there have been many improvements. Such fundamental changes rarely have a single cause. It is in the nature of things that each group involved, knowing its own actions at first hand, believes that it has been the prime mover in development. It may know little about the contribution of other parties. If the IFA is given a central role in this history there is no intention to denigrate the well-meant and effective work of others.

This book does not confine itself to any specialised discipline such as Economics or Sociology. Real life disregards these artificial compartments. People are influenced by history, by economic forces, by society. They are individual; often appearing eccentric to the point of irrationality unless a rounded view is taken of all the influences acting on society.

Most of the early activists are already dead. Some events were never recorded, or the records have disappeared. Quite commonly several versions of an event are current. The authors can give an eyewitness account of most of the development since 1950. They attended many of the meetings described, preparing the minutes and working papers. Their families were involved with the first developments of Irish farm organisation. This book is intended as a tribute to a generation of farmers who took a cool look at their problems and then took constructive action. It was a privilege to work with them.

Louis Smith was employed with the co-operatives from 1950, giving him contacts with Macra and with farm organisations abroad. Recruited as Economic Adviser to Macra six months before the foundation of the NFA, he helped the committees preparing policy documents for the new organisation. After the launch of NFA he continued that work till 1964, combining it with lecturing in University College Dublin. Till retirement he remained in UCD and was Economic Consultant to Associated Irish Banks.

Sean Healy was first employed by the Cork Farmers' Union, then intimately involved with Macra na Feirme, as organiser and as General Secretary, from 1948. In Macra he saw the first beginnings of the ICMSA and then saw the NFA to its inauguration in 1955. He remained as General Secretary of NFA/IFA up to 1985, co-ordinating the internal committees and supervising membership, finances and relations with other organisations at home and abroad.

The book could not have been written without the unstinting help of friends. Michael Ross and Michael-Owen Fogarty cast strict eyes over facts, phrases and syntax. Nevill Chance, Tom Cahill, Anne Gibbons, Dick Keenan and many others gave information. Thanks are due to the IFA for access to their minute-books and other records, in particular to Con Lucey, IFA Economist; to the families of Juan Greene, John Orpen, Jim Kehoe, Myles Smith and others who made their personal papers available. Use has been made of the National and UCD Libraries, National Archives, and newspaper files, particularly of the *Irish Farmers' Journal*. All of these deserve our thanks. The errors are the responsibility of the authors.

1 It all began with small co-operative creameries with sheds for the agricultural store ... An early photograph (courtesy ICOS).

1

The Precursors

ABSENCE OF A STRUCTURED RURAL SOCIETY IN THE 19TH CENTURY

To measure progress you must fix the starting point. Only then can you appreciate the achievement in the present position and the difficulties overcome.

Irish farmers were in a quite exceptional situation in the Penal Days of the 18th and 19th centuries. Ninety per cent of the population belonged to no association apart from the family and the Church. The laws still forbade even primary schools, though illegal 'hedge schools' did fine work. The professions were closed to the vast majority, cutting off leadership.

Societies for the improvement of agriculture served 'improving landlords'. The Dublin Society was formed in 1731, 'to promote technical improvement in agriculture'. Its first Spring Show and Horse Show were held at Leinster House that year. In 1820 it became the Royal Dublin Society. By 1847 agricultural advisers were employed by the RDS, the Veterinary College and the Botanical Gardens, and model farms had been founded. Work was done in technical education, crop and livestock improvement with the aid of subsidy from the Irish Parliament and later from Westminster. When the Department of Agriculture was founded in 1900 it took over most of the work of the RDS and government grants ceased. In 1923 the RDS headquarters moved to Ballsbridge and the former library became the chamber of Dáil Éireann.

The Cork Agricultural Society was founded in 1857. Though it became the Munster Agricultural Society it remained a local body, like the many other county agricultural show societies.

These bodies took no part in policy formation or in representation of ordinary farmers' interests. Local communities were broken by the Great Famine of the 1840s, when population was virtually halved by starvation, disease and emigration.

The first organised struggle of the farmers was to gain security by ownership of their land. Sir Horace Plunkett summarised the development in 1904: 'There is no country in Europe, except the United Kingdom, where the peasant landowners have not some form of corporate existence. In Ireland the transition from landlordism to a peasant proprietary not only does not create any corporate existence among the occupying peasantry but rather deprives them of the slight social coherence which they formerly possessed as tenants of the same landlord.'[1]

The Co-operatives

Victory in the Land War, with the passing of the Land Acts of 1881 and 1885, made rural development possible. Before this time the benefit of any improvement in production or marketing was likely to accrue to the landlord. The new farm owners were in danger of replacing dependence on th landlord with a similar exploitation by the shopkeeper (the 'gombeen man'). Working capital was advanced in return for the crop at harvest. Once tied by debt to a single merchant there was no way the farmer could bargain for a fair price. The dairy farmer was becoming fully dependent on the creamery owner. The co-operatives promoted by Sir Horace Plunkett provided an escape. Starting with the co-operative creamery at Drumcollogher, 1889, they were the first modern farm organisations. The central body, the Irish Agricultural Organisation Society (renamed the Irish Co-operative Society [ICOS], in 1979), was formed in 1894 and the Irish Agricultural Co-operative Wholesale Society in 1897.

The object of the founders was to establish a rural civilisation based on the parish co-operative. Starting with the creamery and its store, the co-operative would provide technical advice and credit, become the centre of social and economic life and represent the farming interest. All the work of Macra na Feirme, farmers' union and the modern co-operatives were to be combined.

Under the leadership of Sir Horace and Fr Tom Finlay SJ, Professor of National Economics at University College Dublin, great progress was made, with 881 societies formed by 1910. The need was greatest in the dairying areas, because of the local monopoly of the creamery owners. Irish share of the English butter market had fallen from 40 per cent in 1848 to 3.3 per cent in 1882 – and the Danish price was 30 per cent higher.

The Irish co-operative development was to some extent premature. Excellent work was done, but the leadership did not come from the ordinary farmer. Reliance had to be placed on clergy, gentry and creamery managers. The life experience of the farmers addressed by Horace Plunkett made them cautious. Most of them were too old to benefit from the national primary schools 1832 – 40 years before similar development in England. In 1850, some 511,000 children attended national schools, the same number as had attended hedge schools in 1824. Most students only reached the second book of the six-book course. They had lived through the Famine as young men. Their introduction to combined action was in the Land League agitation and the Parliamentary Party, with the disillusionment of its Parnellite split. Even in Denmark co-operatives could only be developed by farmers 90 years after they had gained secure ownership of farms and 25 years after the Folkhighschools spread a general education.

The co-operative movement did not grow spontaneously from the grass-roots. The costs of organisation were paid by the wealthy Sir Horace and his friends. From 1888 the salary of R.A. Anderson as organiser was paid by the English consumers' Co-operative Union. The movement was dependent on state subsidies because farmer subscriptions were inadequate.

In the Baltic states co-operatives were used to give practical training and experience to future leaders for the time when independence from Russia could be obtained. In Ireland, Nationalist politicians were at best cool, often openly hostile, both to the co-operatives and to the early farmer organisations. Co-operatives were described as an effort to 'kill Home Rule with kindness'. The movement was weakened because the available leaders were so often landlords and Unionists, like Plunkett, or clergy, rather that working farmers. In the beginning 95 per cent of co-operative chairmen were clergy of one denomination or another. Politicians, then and later, resented the intrusion on their power. Many were trade competitors of co-operatives as merchants or cattle dealers.

Plunkett summarised the political situation of the co-operatives as 'like a dog on a tennis court – cursed by both parties'.

Plunkett was forced out of office in the Department of Agriculture by the Nationalists and denounced by the Unionist Carson for supporting Home Rule. The British armed forces specially selected co-operatives for reprisals in 1920. About 60 suffered total or partial destruction, usually by fire, and some staff were shot.[2] In 1923 Plunkett's house, Kilteragh, was blown up one night by the Republicans. Unsatisfied by their work, they returned the next night and burned what was left. After that he retired to England and moved the Plunkett Foundation library and research centre for co-operative studies from Dublin to London.

After the business recession following World War 1 the co-operatives revived. Creameries were consolidated and made over to co-operatives. Dr Henry Kennedy, who was Secretary of IAOS for 37 years from 1926 onwards, promoted Irish Associated Creameries Ltd (1927) through which 156 creameries bound themselves to market their production. The quality mark 'IAOS' was to be the national brand. In Denmark and later in Britain and elsewhere, such a majority of producers would have merited legal support. The creameries opting out of joint marketing were able to sell without the strict quality controls, while benefiting from the enhanced reputaion of 'Irish' butter and could under-cut selected markets. The scheme failed.

Dr Kennedy, checked in market organisation and in the take-over of the last government-owned creameries, turned to improvement of farming. Particular innovations were the promotion early grazing and silage and better breeding through introduction of artificial insemination.

The co-operative movement concentrated on survival.

Early Farm Unions

The IAOS could not represent farmers as Plunkett had planned. There were few co-operatives in the East and Midlands and so it lacked any mandate to speak on the cattle, pigs or poultry trades. The IAOS was weak in negotiation when over a third of its income came from a yearly government grant subject to annual review.

Farmers made a confusing number of efforts to unite in defence of their interests.

The Farmers' Union was formed in 1916 as a federation of county associations, with a central body and paid Secretary, Paddy Belton, in Dublin. It promoted cattle sales to Britain and negotiated farm wages. With Independence it promoted the Farmers' Party, but expired in the recession and fall in prices after World War 1.

In 1931 the United Farmers' Protection Association was formed to represent all farming interests. Its objectives were limited to agitation for relief of burdens on farming such as land annuities and rates, and obtaining security of tenure of land and a moratorium on seizure of land and stock for debt.

The Land Annuities Association had similar aims. Since 1922 the Free State Government had collected £3 million annuities, payable (for land purchase), to the British Government. With the fall in agricultural prices in 1930 the burden became much heavier, rising from about 6.7 per cent of Gross Agricultural Sales in 1929-30 to 10.1 per cent in 1935. When the new Government of Eamon de Valera refused payment, the British placed taxes of up to 40 per cent on the import of farm produce from Ireland, in order to collect the money they considered due. The Irish Government also tried to collect. Added to the collapse in world prices in the Great Depression the burden on farmers was almost unbearable. To increase the irritation, the land annuities had been remitted in Northern Ireland.

The Treaty of 1938 ended the Economic War with a final payment of £10 million by the Irish Government to settle the land annuities. Trade preferences were exchanged. It was too late. The harm had been done. Ireland had lost its position in the British market. 'So far as I can see the British market is gone forever,' said Eamon de Valera in the Dáil in August, 1933. Farms were impoverished. With the benefit of hindsight we can see that all the strife about the annuities was for nothing in the end. The fall in the value of money reduced the burden of the payments to one thirtieth of their old value, a negligible 0.016 per cent of agricultural income. The Free State Government continued to collect half the old annuity.

Agitation for the abolition of rates was equally unlikely to be successful. Rates rose by an amount almost equal to the reduction in annuities. In 1938/9 rates yielded £6.3 million, of which the rural districts paid £3.2 million. Income tax, surtax and corporation tax combined were only £6.3 million, so no Government could be expected to give up such an important source of revenue. When farmers refused payment, livestock were seized by the Government and auctioned under cover of machine guns mounted on army lorries. At Marsh's auction yard, Cork, two farmers were shot. One died. Seized cattle were branded by shaving the letters 'LAA' (Land Annuities Association). No farmer would bid against the owner at the auction. Cattle were bought back at a nominal price, but a year's milk was lost when the cows were held, unmilked, in the pound.

Times were hard and politics were rough.

Security of Tenure of Land

The Land Commission had power to take land for redistribution to landless men or those with uneconomic holdings. The organised farmers objected to the method of payment. The seller was forced to accept 'land bonds' at face value, though they were worth much less in the market. Consequently, the seller did not get even the low price offered. Farmers were unwilling to invest in land improvement; since no one could be sure that his land would not be acquired compulsorily, the price of land was depressed and considered useless as security for bank loans.

Land Sales for Debt

Farmers had borrowed from banks during the boom of World War 1. When falling agricultural prices during the Economic War made repayment impossible, stock was seized and farms sold to pay the land annuities to the Free State Government or to repay bank mortgages. The land made a poor price.

The Irish Farmers' Federation was formed as an umbrella organisation for all county and special-interest groups after a meeting in the Mansion House, Dublin in 1933. The first conference was held in the Hotel St George, Parnell Square, on 24 September 1936. Fourteen (later 17) county unions represented 7,000 members. These unions kept their separate identities as Cork Farmers' Union, Cavan Farmers' Union etc.. The National Executive met monthly with D. Davis, of Wicklow as President. Elizabeth Bobbett, an Englishwoman by birth but a fervent worker for Irish farmers during 30 years, was General Secretary. Membership was open to all engaged in farming, both owners and workers.

The Irish Farmers' Protection Association amalgamated with the IFF in 1937 and registered as a trade union in April 1937, hoping that it could obtain the favourable position granted to workers in their trade disputes.

Relations with the Government could hardly have been worse. The Federation, though most of its members were also supporters of Fianna Fáil, was directly in conflict with that party in regard to annuities, rates and Land Commission policy. It sought recognition of Agriculture as the key industry of the country; however, Fianna Fáil's industrialisation policy required cheap food. Professor J. Johnston, TCD, estimated that the protection of industry gave farmers a home market of £5 to £6 million a year, but lost exports of £17 million.[3] There was little common ground for agreement with the Government.

At the start of World War 2 the Taoiseach, Eamon de Valera, and the Minister for Agriculture, Dr Ryan, received a deputation of IFF. The deputation reported that every one of their suggestions on agricultural policy had been rejected. Dr Ryan warned them that the safeguards given to workers' unions did not apply to the Federation. Whether registered as a trade union or not, no agricultural structure similar to the trade unions in bargaining or consultative status could be contemplated by the Government.

Strike action was taken from 8 November to 1 December 1939. Many fairs and markets were cancelled because of the farmers' boycott, and milk to the Dublin market was almost cut off.

After the War international contacts were needed to secure export outlets for agricultural products. The IFF joined with Cork Farmers' Union and commodity associations to found the Irish Primary Producers' Association as an overall body to qualify for membership of Confederation Europèenne de l'Agriculture. Their application was not successful.

Friendly contacts were made across the water with the National Farmers' Union of England and Wales and the Farmers' Action Council. The Devonshire Farmers Association came on a fraternal visit in 1947. The Scottish Farmers' Union refused to open relations; 'contacts were better made through the normal government agencies'.[4] The last general meeting of the IFF was held in 1960.

Commodity Organisations

Special interest groups were formed as the need arose – usually when scattered producers had to negotiate with a monopolist trader or processor, or for joint marketing of a product. The oldest commodity group was the Federation of Beekeepers' Associations, formed in 1881. South Leitrim Beekeepers' Association combined with 17 groups from other areas to form a national association in 1902. It had 1,600 members in 50 associations.

The Beet Growers' Association (BGA) was formed in 1926 to negotiate beet prices with the newly established sugar factory in Carlow. The Act setting up the industry gave the BGA the right to meet its expenses from a levy on all beet sold. In effect membership was compulsory for all beet growers. Because its members were tillage farmers it extended its work to cover all grain and vegetable crops as they expanded under the drive to self-sufficiency in the 1930s. It negotiated with the Flour Millers' Association on wheat and with Arthur Guinness on malting barley. A feature of the BGA was the active involvement of politicians of various parties, typified by Martin Corry TD (Fianna Fáil), its long-term and most forceful chairman. John Greene, was a founder member and chairman of the BGA. The BGA claimed 7,000 members at the outbreak of World War 2.

The Pig Breeders' and Feeders' Association was founded in Cork, under the auspices of the Cork Farmers', to prepare evidence for the Pig Industry Tribunal of 1933. From Cork it spread to other areas and claimed 7,000 members in 1940.

Associations of milk producers figure prominently in this history. The producers of milk for direct human consumption were the first to organise.[5] Before

Pasteurisation and refrigeration the product was so perishable that deliveries had to be made twice daily. Production was in the immediate neighbourhood or from 'flying herds' milked within the town itself. There were over 200 such byres in Dublin city, some with over 100 cows. A few farmyards remained in the built-up area even to the 1950s. The Cow-Keepers' Association of producer-retailers needed little organisation. The milk producer was a small-scale trader, a towns-man rather than a farmer. One of them even became lord mayor of Dublin.

Gradually the system changed. The retailers became wholesalers. Pasteuri-sation, introduced in 1924, widened the catchment area, especially when milk lorries taking 30 twenty-gallon milk cans at 30 mph replaced horse-drawn drays. Now the producer was a farmer, without direct contact to the consumer. He was fully dependent on the wholesaler. During the summer surplus he took what price he could get. Often the wholesaler would leave milk by the road, uncollected, once the lorry was full.

The consumer was also dissatisfied. Doctors warned against the dangers of dirty milk. The Milk and Dairies Act 1935 and subsequent regulations set rules for hygiene and the licensing of producers, wholesalers and retailers. These regulations, up to date in 1935, were never improved and there was no incentive to supply anything above the minimum quality.

Cork Liquid Milk Producers formed their organisation in 1934. In the Dublin area county organisations began to form in 1935. A week long strike in July 1936 was successful, obtaining an increase of 25 per cent. More important was the Milk (Regulation of Supply and Price) Act which set up the Dublin and Cork Milk Boards to regulate city markets. A monopoly of supply was given to the farmers on the register of milk producers in the traditional catchment areas. Board members were elected by the producers, retailers and wholesalers. Though the powers were limited (it could not fix prices and the minister appointed the chairman) this was a great step forward.

The Act made it essential to form an association to represent producers on the milk boards. County associations supplying Dublin federated on 12 May 1937, forming the Leinster Milk Producers' Association. Their candidates have dominated all the elections since then. Being in close contact with urban workers, farmers copied the trade union seen to be successful in the neighbouring cities. The wholesalers were seen as the employers; the registered producers as the workers. The right to supply milk to Dublin was restricted by law to a certain area and to licensed farms. Possession of a 'quota' added to the value of a farm. This made a 'closed shop' possible. Milk could only be bought from association members. Farmers who cut prices were prosecuted and fined under the powers given to the Board. Strikes were seen as the farmers' weapon, just as they were the workers'.

A Chaos of Organisations

A multitude of organisations flourished and asserted independence. James Dillon said he had to consult 63 organisations to get the farmers' view.[6] The NFA counted 97. A lesson was learned when Canon Hayes of Muintir na Tíre led Tipperary farmers to the Minister for Agriculture asking for higher grain prices and met a deputation of Cavan pig producers, come to ask for a reduction in feed prices.

During and immediately after World War 2 little attention was given to co-operatives and almost none to the need of farmer representation outside the farm gate. The Committee on Post-Emergency Agricultural Policy in all its reports had little on co-operatives and never mentioned the subject of farmers' unions or associations. The Commission on Vocational Organisation (1943) had two paragraphs and no recommendation on farm unions.

In the *Irish Independent*'s booklet 'Post-War Planning in Irish Agriculture' there were ten contributors, of whom two made reference to the need for organisation. J.N. Greene, ex-chairman of the Beet Grower's Association, pointed out: 'When the Dublin Cattle Market removal was being discussed no farm organisation was in the list of those consulted. Though I am criticising the procedure I can, to a large extent, sympathise with the authorities because, as a matter of fact, there is no really representative farmers' organisation to speak on all problems for the farmers of the entire country ... There have , instead, been too many self-constituted groups who claimed a hearing without really representing any but themselves. To satisfy this urgent need it is, in my opinion, necessary that the Government should encourage the formation of a farmers' union that will be really representative of all the farmers' interests.' Senator E.R. Orpen wrote: 'In this country, where the bulk of the wealth comes from agriculture, where half of the working population is engaged in agriculture, where there are numerous specialised agricultural societies, yet there is no recognised agricultural group accepted as speaking for agriculture as a whole. If there were a group or organisation, many of the difficulties which now confront the Department of Agriculture might be dealt with in a far more satisfactory manner by the farmers themselves.' (John Greene was father of Dr Juan Greene, first President of NFA. Senator Orpen was father of John Orpen of NFA Grain Committee. It is remarkable how the same families provided leaders with similar views in several generations.)

FARMERS IN POLITICS

There was no room for farmers in party politics before Independence. The Nationalists were united in seeking independence but had no agreed agenda in social or business policy. The Sinn Féin of 1917 to 1922 was the political face of the armed struggle for independence. With independence and the Civil War it dissolved. In 1922 there were 1,600 branches; in 1923 only 60.

The two main political parties which emerged with independence, Cumman na nGaedheal, now Fine Gael, and Fianna Fáil, are based on conflict in the Civil War with no basic difference in social or economic policy. The Labour Party has usually drawn only 10 per cent of votes. There was not, therefore, any party with which it was natural for farmers to align. Efforts to establish farmers' parties in 1923 and again in the 1930s had short term successes. Even if the members had agreed on farm policy, they had no unity on other political issues. They lacked a unifying philosophy or national objective. Their attitude to development was negative, seeing, for example, no need for a Department of External Affairs or the ESB.

In 1923 the Farmers' Party, the political wing of the Farmers' Union, was confident of playing an important part in the new Free State. They put forward 64 candidates but won only 15 seats, though farming was the livelihood of over half the people. No seats were won in the small farm areas of Connaught, Kerry or Monaghan. The more prosperous farmers sided with Cosgrave's Cumann na nGhaedheal Government as most likely to restore order; promote agricultural trade; increase production and cut input costs.

By 1927 some of the farmers were becoming attracted by Fianna Fáil's protectionist policy. However, the Farmers' Party leaders, F.J. O'Hanlon (Cavan) and T. Gorey, favoured fusion with Cumann na Gaedheal, 'to make the largest party in the state overwhelmingly agricultural.' The party decided to remain independent, while supporting the Government.

The real contestants for government were the two parts of Sinn Féin, Fianna Fáil and Cumann na nGaedheal. The Farmer's Party declined in each election. Failing to attract the small farmers of the West, who remained firmly Fianna Fáil, it came to represent the larger farmers.

As the Farmer's Party declined, moves started (1931) to form a wider based organisation, the Agricultural League, advocating a Central Bank, salary reductions in the public service, derating of land. They did not contest the 1932 election, but supported three rural Independents who were elected – James Dillon, J.F. O'Hanlon and F. MacDermot. They opposed the Fianna Fáil declaration of the Economic War. Shortly after the 1932 election the farmers formed the National Farmers' and Ratepayers' League When James Dillon joined in 1933 the party was renamed the Centre Party.

In the 1933 election the party won 11 seats. Their platform was ending the Economic War, friendship with Northern Ireland and an end to civil war strife. It became the third largest party, in spite of the disruption of its meetings by Fianna Fáil and IRA supporters. It is hard to believe the accounts of the bitterness of politics in the aftermath of the civil war; the degree of interference with freedom of speech by both sides and the way in which business was irrelevant to politics.

The Centre Party lasted only a year. In 1933 it merged with the National League (Redmondite) group and Cumann na nGaedheal to form Fine Gael. While seven of the individual farmer members were elected in 1937 the farmers had no formal representation in the Dáil. All politics was focused on past bitterness.

The small farmers, till now firm supporters of Fianna Fáil, became discontented in the later 1930s. The protection of the grain growers raised the price of feed, which almost destroyed pig and poultry production in the West and North. Clann na Talmhan was formed at Athenry in 1938, led by Michael Donnellan. It stood for the small farmer, support for land reclamation, tax relief and afforestation. For 20 years from 1943 it played an active part in the Dáil, with a maximum of 10 seats, aided by about four independent farmer TDs. Michael Donnellan (father of John Donnellan, Fine Gael TD, Galway) and Joe Blowick of Mayo held office in the coalition governments of 1948 and 1954, but, once again, the voters faded into the national parties of Fine Gael or Fianna Fáil.

Farmers were under represented among TDs. Twenty-two per cent of deputies gave farming as their occupation compared with 35 per cent of those gainfully employed in 1961. They were divided between Fianna Fáil and Fine Gael with a slight preference for Fianna Fáil. These people were not elected as farmer representatives or through membership of farm associations. The route to election was through local politics (56 per cent), family relationship (23 per cent), or prominence in sport (13 per cent).[7] There was no farm lobby.

Many progressive farmers echoed William Butler Yeats:

A statesman is an easy man,
He tells his lies by rote,
A journalist makes up his lies
And takes you by the throat;
So stay at home and drink your beer
And let the neighbours vote.

('The Old Stone Cross.')

It was in this spirit that Macra refused James Dillon's offer of a seat in the Seanad. They feared the divisive influence of party politics outweighing the association's interests. The NFA followed the Macra rule that any member taking a political post must resign any local or national office held in the association. This covered the case of farmers co-opted to County Committees of Agriculture under the Local Government Act (1964). James Dillon himself, strongly recommending the formation of a farmers union, warned the founders of NFA to exclude all members of the Oireachtas from office because political loyalties would out-weigh those to farming.[8]

Table 1 Seats Gained by Farmers' Parties in the Dáil

Party	*Election*	*Representatives*
Farmers' Party	1923	15
	1927	11
	1927	6
Centre Party	1932	11
No representation	1933 & 1937	
Clann na Talmhan	1943	14
	1944	11
	1948	7
	1951	6
	1954	5
	1957	3
	1961	2
	1965	1

[Source: Mair P., *The Changing Irish Party System*, London 1988]

RELATIONS WITH GOVERNMENT

The Department of Agriculture

Plunkett called together politicians from all parties in the Recess Committee (1896) to hear evidence and to recommend improvement in Irish agriculture. Its unanimous report was that the Government should establish a Department of Agriculture and Technical Instruction. It should act in conjunction with farmers, through their organisations: 'It is everywhere on the Continent now recognised as a principle – first, that the action of the people themselves, through industrial combination, is more important than the action of the State; and, secondly, that the assistance of the State can only be truly effective when there exists a system of local representative organisations to co-operate in its administration. All attempts of the central administration to act through unorganised individuals, in schemes of agricultural and industrial improvement, are by implication condemned as likely to do more harm than good. We need not enlarge on this lesson; it is enforced with emphasis throughout the entire series of our foreign reports, and in the memorandum of M.E. Tisserand (French Director General of Agriculture). For Ireland this lesson seems more vital than it is for other countries longer inured to habits of industrial enterprise. The effects of organisation on character are even more valuable than its economic advantages. It engenders self reliance and mutual confidence among the people; it sharpens their intelligence, systematises their habits, and opens out to them new conceptions of their own powers and of the resources of self help. Without organisation spreading pari passu among the people, state aid is in danger of stepping beyond its proper limits and may prove distinctly mischievous. A lavish expenditure of money, which taught the people to lean more than ever upon the government, and still further weakened their backbone, which demoralised officials and the public by wastefulness and jobbery, would leave the last state of the country worse than the first.'[9]

Plunkett, introducing the Act in the House of Commons, made the Government's intention clear: 'An agricultural department is now a remedy, and as long as its functions are limited to what such departments have done successfully elsewhere it should certainly be created. But that any government assistance, whether legislative or administrative, can do one tithe of the good which can be done by organisation, is a popular fallacy.'[10]

Department and Co-operatives

When the Department of Agriculture and Technical Instruction opened on 1 April 1900 it was, on the unanimous recommendation of the Recess Committee and confirmed in its enabling Act, to work with and through farm organisations. There was to be a Council of Agriculture which 'shall meet at least once a year for the purpose of discussing matters of public interest in connection with any of the purposes of this Act'.[11] The Council of Agriculture shall consist of: 'A number of persons resident in each province equal to the number of counties, with due regard to the representation on the council of any agricultural or industrial organisations in the province ... The members representing each province shall constitute separate committees on the Council and shall be styled the provincial committees of the respective provinces.'[12] The Department was to work 'in such a manner as to stimulate and strengthen the self-reliance of the people' through co-operatives, the Boards of Agriculture and of Technical Instruction and the county councils.[13]

Sir Horace was appointed first head of the Department to ensure this co-operation with organised farmers. He took IAOS staff with him to provide continuity and close contact.

From the first, relationships were unhappy between Plunkett, Department and IAOS. Nationalist politicians and the trade interests were determined to remove Plunkett. They objected to his politics and his co-operatives. The new civil servants exercised authority over their former colleagues in the IAOS, leading to personality clashes.

In the beginning the Department contributed substantially to IAOS funds and Plunkett made over his salary as vice-president of the Department to IAOS. In May 1906 the Department's Council of Agriculture, which had a large measure of control of departmental funds, resolved: 'That, in the opinion of the Council, it is desirable that the Department should promote agricultural organisation and provide the funds necessary for the purpose.' Six months later, in November, the Council insisted that 'the IAOS, which is the only existing body having special knowledge of this work' should be aided.

Plunkett was removed from his position in the Department in 1907 in response to pressure from private traders and the Nationalist Party, whose leaders John Redmond and John Dillon (father of James Dillon, the future Minister for Agriculture) made strong attacks on Plunkett and the co-operatives in the House of Commons. In particular they attacked the 'controversial co-operatives', that is, those competing with their private commercial interests.

Under T.W. Russell, Plunkett's successor as vice-president, the new Department took a course opposite to that laid down by the Recess Committee and by the Act under which it was formed. Dan Hoctor, a most devoted official of the Department, describes Russell: 'He was unorthodox in his Conservatism. He turned to Liberalism early in the new century and he ultimately became a Home Ruler. A man of strong character, who never pulled his punches in debate.' Tim Healy, who admired his abilities, said of him: 'Devoid of the geniality and humour of his race (Russell was a Scot), he sported a bilious face and splenatic manners. Russell's dislike of the IAOS was notorious; his conviction was that the responsibility for organising agricultural co-operation should rest directly with the Department. His agricultural policy did not differ much from his predecessor, except with regard to the IAOS, and he continued to build on the foundations already laid.'[14]

Russell's appointment was the most important in the history of Irish agriculture. He set the pattern of the Department, its breeding policy for livestock, its relation with farm organisations, for the next 60 years.

Co-operative representatives and sympathisers were summarily removed from the Council of Agriculture by Russell. Some men appointed to the Council 'had not enough land to sod a lark'. The resolution of 1906 on aid to co-operatives was rescinded. The IAOS was notified that the subsidy (£3,700 when total income was £4,100) would be phased out over three years. Meanwhile, its work should be confined to organising and auditing. It was forbidden to give technical advice. Salary scales and expenses were subject to Department approval. No co-operative was to be organised without Department approval. The co-operative newspaper, the *Irish Homestead*, had to be separated from the IAOS. Anderson, IAOS Secretary, wrote: 'I found it very hard to control the frayed tempers of my small staff. The monthly expenses and diaries of organisers, my own included, were subjected to rigorous examination and criticism, so that much of our time was taken up by why a telegramme rather than a letter was sent or why an organiser hired a car when there was a railway (but no suitable train!). The whole ... staff flared up when it was discovered that an official on £1,000 salary and first class expenses had dogged the itinerary of one of our most trustworthy and zealous organisers whose humble pay was £250.[15] The Department supervision of salaries, expenses, hours of work was still very much in evidence when Louis Smith worked with the IAOS in the early 1950s.

The IAOS subsidy was cut off, without further warning, in 1908.

In a way, the removal of the subsidy benefited the co-operatives. Public support and increased affiliation fees from the members made good the lost income. Well-wishers presented No. 84 Merrion Square, to Plunkett, as head offices. By 1919 the IAOS, with 100,000 members, was the largest agricultural association in the British Empire, a model copied in Britain, USA, Finland and elsewhere.

Credit co-operatives prospered in the early years of the century, numbering 268 in 1908. T.W. Russell was a strong critic of this development, opposing the aids given to them at their foundation. The Department of Agriculture grant to IAOS for organisation and supervision was withdrawn in 1908. In 1910 Russell opposed the Congested Districts Board grant of £350 to IAOS for organising, declaring that the loans 'were not worth 13 pence in the pound', and called for the refunding of the seed capital advanced to the societies. All but £98 was repaid to the Government, but the credit societies were drained of capital. The Department Committee on Agricultural Credit, 1912, found that the credit societies had been most beneficial to farmers and recommended that the Deparment set up a special section to aid them. This was not done and the credit movement almost died.[16] The significance of this episode is in the contrast with other countries, such as Finland, where seed capital from the Government enabled the societies to take off to self-financing, farmer-controlled banks – the course recommended again for Ireland in the Gilmore Report of 1959. (See Chapter 6.)

The IAOS subsidy was renewed in 1913. Financial dependence on the Department (38 per cent of IAOS income in 1938) continued to be a means of control. In 1922 the Department of Agriculture in the new Irish Free State insisted that the 32-county IAOS confine its activity to the 26 counties. The Ulster Agricultural Organisation Society had to split off, – to the great loss of the movement. The wholesale society, IAWS, not being dependent on state grant, continued to serve the 32 counties. Margaret Digby, in her foreword to the UAOS history,[17] describes relations with the Ministry as similar to those in the Free State: 'Most agricultural co-operative movements, whether in Europe or America, can count on the active support of their Ministries of Agriculture. In Northern Ireland support came only intermittently: a push at the beginning followed by a long period of discouragement ... Marketing Boards and Government controls of farm produce, which in most countries are built around the farmers' own marketing organisations, here seem to have quite another basis. There were years when one door after another was not so much closed to the Ulster farmers' co-operatives, as slammed in their faces.'

Government and Farm Organisations after Independence

Relations between government and farm organisations in Ireland were out of step with the rest of Europe.[18]

The source of difference is traceable in part to clashing personalities but more especially to the British institutions which, as de Tocqueville observed in his travels,[19] had effects in Ireland quite contrary to those in England itself. The Department became dominant because it controlled finance and technical knowledge as well as administration. The Agricultural Councils and the Agricultural Board, created under the 1900 Act and with some financial responsibility, disappeared with the achievement of national independence in 1923.

In practice the Ministers for Agriculture have usually been politicians without farming connections. Several have had business interests in conflict with the current organisational efforts of the farmers. In 1953 James Dillon, Minister for Agriculture, spoke to Dr Henry Kennedy, Secretary of the IAOS, on the telephone, in the presence of Louis Smith, stating that the IAOS subvention was given to promote co-operatives processing and marketing agricultural produce. He should restrain the said Louis Smith from organising societies supplying goods. They were competing unfairly with private traders. (Some small societies had been started in county Mayo, not far from Ballaghadereen, where the Minister was a merchant.) Dr Kennedy replied, 'Is that a threat?' and directed that the work of organisation continue – but to remember what it might cost.

This system, so contrary to the structures of Denmark, Netherlands or France or to the practice of the British, with their Agricultural Marketing Acts, was to lead to bitter confrontations in the early years of NFA.

The first, Cumann na nGaedheal, government had excellent relations with the IAOS. Dr Henry Kennedy, Secretary of IAOS, was brother-in-law to the Minister for Agriculture, Paddy Hogan. A much needed restructuring of the dairy industry took place. All creameries were bought by the Government through the Dairy Disposal Company. This company was to consolidate creameries into viable units and sell them back to the co-operatives. In some poorer areas and where the Civil War made co-operation between neighbours impossible, the company retained ownership.

In Ireland each service was centralised under political control. Farm advisory services were under the county committees and the Department; the boards of state companies, such as the Sugar Company or Dairy Disposal Company, were politically appointed nominees. Even state companies were not involved in any structural way in the formation of policy.

County Comittees of Agriculture

County Committees of Agriculture, established in most counties after 1907, administered various Departmental schemes. The members were appointed by the county councils to reflect party political, rather than local or farming interests. In some cases few, if any, of the members were connected with farming. The prime interest of many of the county committeemen was in party politics or in the few 'percs', such as travel expenses, to be obtained. Even in 1976 Sligo County Committee of Agriculture called for the co-operatives to be taxed – hardly in the interest of farmers.[20]

The County Committees of Agriculture could have been, like the French Chambres de l'Agriculture, natural representatives of farmers, active in developing marketing co-operatives. The agricultural advisory officers appointed under these committees were instructed that their functions were restricted to technical advice on production. The education of these advisors at the Albert College and UCD was funded by the Department and confined inside the farm gate. The post of Professor of Agricultural Economics was created in the Act of 1926 but unfilled till the appointment of Professor Seamus Sheehy in 1972. In earlier times 16 lectures, covering all aspects of economics and marketing (given by Louis Smith) were considered adequate for the formation of future agricultural graduates.

In practice the few instructors who ignored this restriction, such as John Murphy, CAO Wexford, in fruit marketing, had no action taken against them and performed a great service.

The farm organisations saw no practical advantage in representation on these County Committees.

Consultative Councils

The Agricultural Act (1931) gave power to the Minister to form consultative councils for commodities. Eight councils were formed: Dairy Produce(0), Livestock (1), Eggs (7), Tobacco (1), Wartime Production (8), Poultry and Rabbits, Milk (Dublin District). (Figures in brackets show the number of meetings held in three years prior to 1941.) There were no standing committees or staff. These councils were appointed by the Minister, who presided himself or appointed a chairman. They had no responsibility to farmers. They only met at the Minister's infrequent invitation. Even then they were debarred from discussion of any matter not contained in the agenda prepared by the Minister. The Minister did not think it possible to discuss proposed legislation in depth because it was the practice not to divulge the text of Bills before they were presented to the Oireachtas. Consequently any advice could only be in the most general terms.[21] Such councils could develop no continuity and were of little significance except as an indication of a governmental attitude. They prefigured Neil Blaney's National Agricultural Council of the 1960s.

The Government Commission on Vocational Organisation summarised its findings:

> In the first place, it appears that agricultural planning, in so far as it exists, is in the hands of the officials of the Deparment of Agriculture, not of the farmers.
>
> In the second place, apart from the Agricultural Production Council, established owing to emergency conditions, the Department is not assisted or advised in the discharge of its functions by any general council or board such as those which existed before 1922.

> In the third place the arrangements made for consultation of particular interests by means of sectional consultative councils cannot be regarded as giving agriculturalists concerned an adequate voice in the determination of policy or the supervision of administration.
>
> Finally, it is apparent that there is little collaboration between the Department and organised agriculturalists. Broadly speaking agriculture is regulated on bureaucratic lines and the administration of the Department's regulations and orders necessitates the employment of numerous officials and inspectors.[22]

This was, of course, not a policy confined to agriculture. The Government considered itself the only body able to handle many problems in the new state. The state service expanded to fill a vacuum. Unlike the clergy, who provided leadership in the early co-operatives, the central Government did not yield responsibility to vocational groups as they emerged and showed competence. Centralised state control increased in farming, industry and in areas such as education and local government.

An independent report to the Government stated in 1952: 'The fact that Ireland, a country that regards itself as nonsocialistic, should adopt a pattern of industry, under which the proportion of its investment resources channelled through Government initiative is almost twice that which was adopted in the United Kingdom under its Labour Government, has a far greater significance than merely the presenting of an interesting paradox.'[23]

Government Reaction to Farm Agitation

Government control of agriculture and the prices of produce led naturally to conflict. Asserting control of everything it was blamed for everything – even for events impossible to control. Trade union law protected the workers' right to strike. Farmers, being self-employed, had no legitimate way to make forceful protest. They were not consulted on matters which affected their livelihood and came into conflict with the law if they placed pickets or withheld their produce from sale. Other governments took a contrary position. Britain under the Marketing Acts of the 1930s and the USA with the Kapper/Volstead Act, made special provision to enable farmers to control the marketing of their products so as to gain equality in marketing strength with other sectors of the community.

Strong measures were taken against farmers' associations. Imprisonment was repeatedly used, beginning with the land agitations of the last century; again in the 1930s, when one farmer was shot dead and another wounded during a sale of cattle seized during the Economic War.

The Federated Farmers met with the Liquid Milk Producers to co-ordinate action during 1938. A price concession for the nine off season months averted the strike, but in 1940 the Milk Board recommended a price increase of 1.7 pence a gallon. Instead, the Minister reduced it by 1 penny (20 per cent). At the same time statutory agricultural wages were raised and the price of farm supplies had been decontrolled, allowing sharp increases in production costs.

Leinster Milk Producers sought registration as a trade union, hoping to gain immunity for pickets on milk processors. Dr Ryan, Minister for Agriculture pointed out to them that this would serve no purpose. Farmers, rich or poor, were self employed. They had no employer against whom they could legitimately picket. Any agitation to withhold supplies was a 'conspiracy in restraint of trade' and subject to prosecution.

The milk suppliers responded by a milk strike lasting from 5 November 1939 to 6 January 1940. The Farmers' Federation organised a boycott of fairs and markets.

Dr Ryan reacted strongly against what he considered a political threat: 'If you ask me who are the Farmers' Federation, I can tell you that there may be a few ex-members of our organisation there, but the driving force behind it is the same that has oppressed you and me in every movement we have taken part in since we inaugurated the Irish Volunteers in 1913. The same force will be against us to the end. They will also oppose any government that stands for the national rights of the country.'[24] These were strong words to use about an application for an increase in the price of milk.

Each group of politicians in turn tried to politicise farm organisation, as John Redmond and John Dillon had branded the co-operatives 'Unionist' at the turn of the century, each politician interpreted social and economic unrest in terms of the irrelevant politics of the time.

The Offences against the State Act was used against the picketing farmers during the milk strike, accused of intimidation and interference with produce and property in the milk strike of 1940. Eighty well known farmers were arrested and sent to jail. The Censor banned all reports of strikes or meetings in the newspapers and radio for reasons of national security.

In the Dáil Dr O'Higgins moved a motion disapproving the Government's use of the Act: 'The only step taken by the Government was to brand these people as criminals and bring them before the Military Tribunal. That was recklessly provocative action calculated to make the strike a strike of desperation ... We were told that the powers were necessary to deal with armed conflict and rebellion. Yet in the first week of the farmers' strike the Special Powers were put in force against the farming community. The censorship of news and views were driving people to ferocity. In some counties the first weapon was censorship, the second the use of military force, and the third the power to seize goods.'

Milk was requisitioned from creameries. Those which complied were given a permanent right to supply the Dublin market; the 37 co-operatives which refused were fined £100 for the first day's offence and £10 for each succeeding day – a heavy fine when manufacturing milk was two pence halfpenny a gallon and cows cost £16 a head. The fines were reduced on appeal. Ending the strike farmers gained a right to yearly contracts and the Minister's promise 'to consider' a price increase. The cost to farmers in lost income was about £200,000 (in 1996 terms about £6.6 million).

Manufacturing Milk Strike, 1952

A suppliers' strike was called in February 1952 which lasted 16 days. The proposed price cut was averted and an increase of 1.25 pence a gallon obtained. In the climate of the time this was a major victory for militancy. The ICMSA met the same resistance to recognition by the Government as the Federation had experienced in the past and the NFA would do in the future. James Dillon made this clear in the Dail debate on the increase in milk price.[25]

Milk Strike, 1953

The end of the War put pressure of farm incomes. Between 1948 and 1953 government-controlled farm wages rose 40 per cent and uncontrolled fodder by 55 per cent, while the fixed milk price rose only 9.5 per cent. The Milk Act of 1936 had provided that milk prices be fixed at 'an equitable level' without giving any indication of how this should be decided. The Government appointed a Milk Costings Commission in 1951 but it was so slow in reporting that the LMP hired its own experts to cost production. In any case the question was not one of business but of politics. The Department of Industry and Commerce allowed milk wholesalers an increased margin to cover costs; the Department of Agriculture refused to meet farmers to discuss prices, stating in a letter that 'no useful purpose would be served'.

The suppliers to Cork and Dublin combined with the Irish Creamery Milk Suppliers' Association (formed in 1950) to restrict the milk supply to cities from 24 December 1952 and creameries withheld their supply to Dublin on 22 January 1953. Stoppages occurred in towns through the country in support of the demand for 4 pence a gallon increase for liquid milk and 2.5 pence for manufacturing.

The suppliers' position was weakened when the producer/retailers refused to support the strike and continued to supply 18,000 gallons a day to Dublin. A new development was the importation of dried milk from the Netherlands for reconstitution, giving consumers 50 per cent of normal supplies.

Controversy was strong. The farmers' behaviour was denounced as 'unIrish' – though 174 urban strikes in the same year were not condemned. The Minister for Local Government, Paddy Smith, called the farmers 'Blue Shirts', to which Pat Ryan, Secretary of the LMP, replied that his members could only afford hair shirts. Ninety-four farmers were jailed.

The Minister for Agriculture, Tom Walsh, refused mediation; the Taoiseach, Eamon de Valera, refused to make any decision on prices till supplies were resumed. The farmers would not accept this defeat. Eventually de Valera wrote to the Housewives' Association suggesting an increase of 0.83 pence. On 11

February the LMP decided to take this as a face-saving offer and called off the strike.

The ICMSA staged a march of 18,000 through Dublin on 7 March 1956[26] The ICMSA campaign began before that of the NFA. The high point was the illegal picketing of the Dáil while it was in session. Farmers were available in relays to take up position and be arrested till 452 had been taken to the garda barracks. They did not serve prison sentences as their fines were mitigated to £1 each and a sum of £448 was paid by the Leinster Milk Producers. The milk price was raised by two pence a gallon. This agitation, successful though illegal, impressed all farmers.

CONCLUSION

Irish farmers had failed to unite. Efforts had been made in co-operatives, farmers' unions, commodity associations and political parties without much success. Their views were ignored on the conduct of policy on the farm; on breeds of stock; on marketing; on pricing policy. Had the national policies adopted been even moderately successful all might have been well, but the results were disastrous.

The Department of Agriculture, set up to work for farmers through their organisations, had come to be seen as alien and hostile to farm interests.

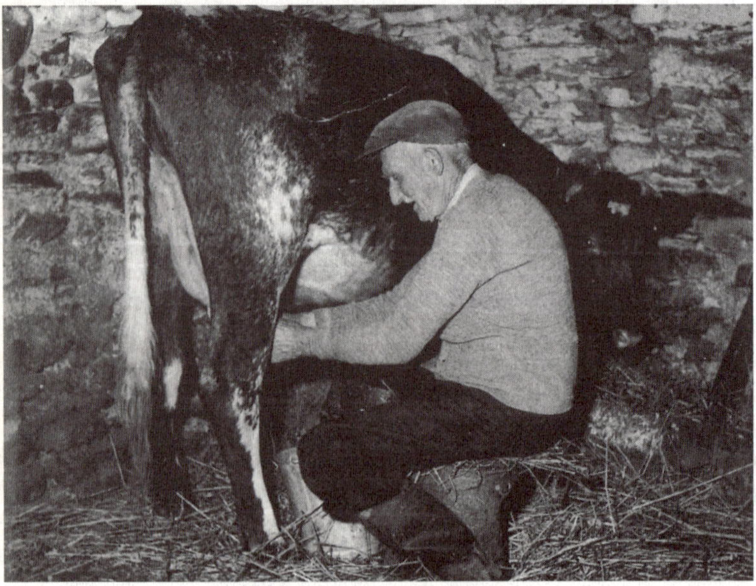

2 Hand milking

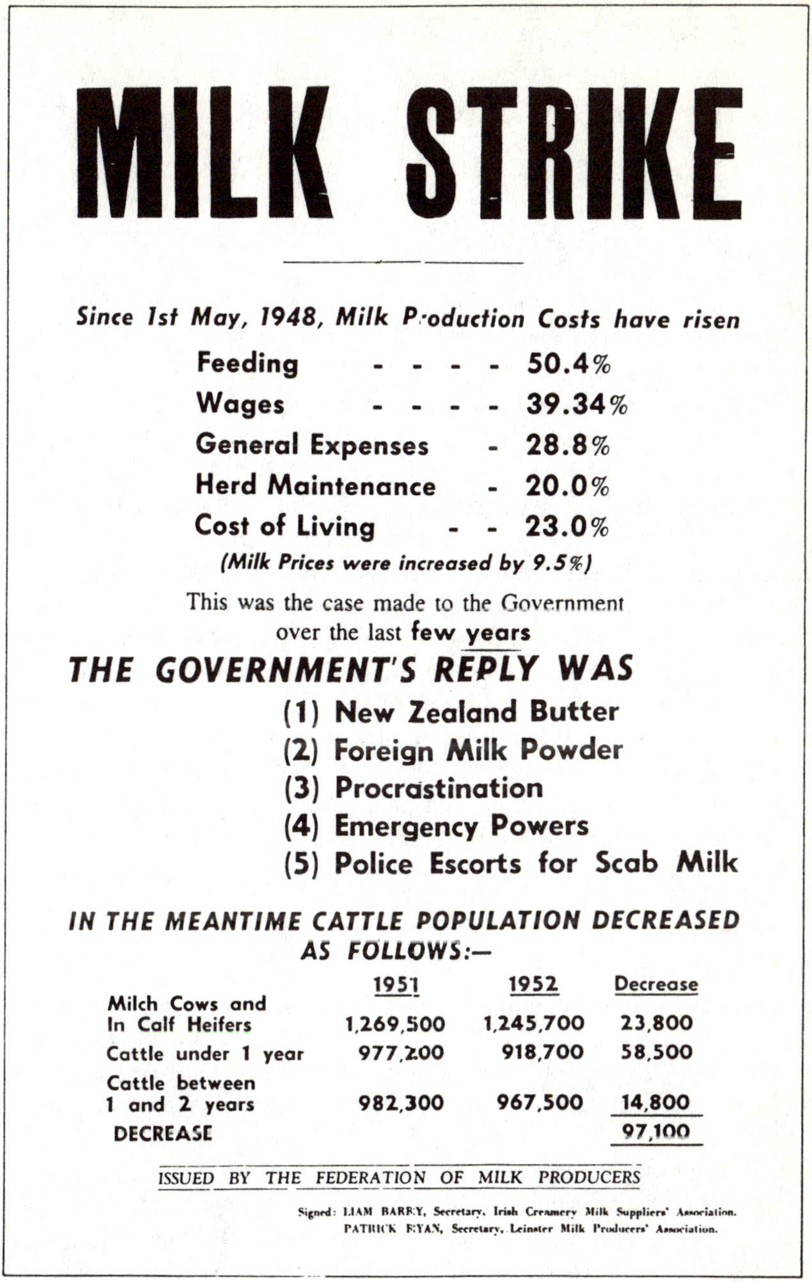

MILK STRIKE

Since 1st May, 1948, Milk Production Costs have risen

Feeding - - - - -	**50.4%**
Wages - - - - -	**39.34%**
General Expenses -	**28.8%**
Herd Maintenance -	**20.0%**
Cost of Living - -	**23.0%**

(Milk Prices were increased by 9.5%)

This was the case made to the Government over the last **few years**

THE GOVERNMENT'S REPLY WAS

(1) New Zealand Butter
(2) Foreign Milk Powder
(3) Procrastination
(4) Emergency Powers
(5) Police Escorts for Scab Milk

IN THE MEANTIME CATTLE POPULATION DECREASED AS FOLLOWS:—

	1951	1952	Decrease
Milch Cows and In Calf Heifers	1,269,500	1,245,700	23,800
Cattle under 1 year	977,200	918,700	58,500
Cattle between 1 and 2 years	982,300	967,500	14,800
DECREASE			97,100

ISSUED BY THE FEDERATION OF MILK PRODUCERS

Signed: LIAM BARRY, Secretary, Irish Creamery Milk Suppliers' Association.
PATRICK RYAN, Secretary, Leinster Milk Producers' Association.

3 The farmers' case, 1953

2

The State of Agriculture, 1945–1955

Ireland may have been an agricultural country after World War 2 but Irish agriculture was in a sorry state. The land was starved. Compulsory Tillage Orders in 1944 compelled farmers to till 38 per cent of land loosely defined as 'arable' and often quite unsuited to wheat. Repeated cropping without fertilisers (only 1,700 tons were imported for the 26 counties in 1942) combined with bad weather to reduce yields of wheat from 2,500 Kg per Ha in 1940 to 1,355 in 1947. Yield reached 7,900 Kg per Ha in 1992. Cattle numbers had been static for 50 years. Professor Murphy, UCC, found an average yield of 460 gallons, after deduction milk fed to calves, in the Golden Vale in 1938 and 426 gallons in 1942.[1] Output was 100 gallons an acre, and declining, compared with 400, and rising, in the Netherlands. Naturally, yields were much lower in areas of Ireland with poorer land and harsher climate. National output was 240 million gallons – the supply to one major creamery in 1995. Five thousand tons of butter had to be imported in 1953. The export of wild rabbits (4,800 tons) was nearly twice that of mutton and lamb. Protection of tillage excluded the cheap grain available to competitors. Faced with penal tariffs in the Economic War and discouraging prices in the World War, exports of intensive products, such as bacon and poultry, had fallen. On one sixth the area, Northern Ireland produced as many pigs, twice as many poultry and 80 per cent as many tractors as the Republic.

Foreign Trade

Foreign trade directly concerns Irish farmers. In 1955 33 per cent of production was exported. The home market was already satisfied, so any increased output had to be exported. In fact the proportion exported rose to 64 per cent in 1980 and 75 per cent in 1993, but there was no co-ordinated plan for marketing.

Trade Relations with Britain

Folk memory of international trade policy was unhappy. It was not business oriented. During the Economic War British tariffs of 40 per cent cut the value of Irish agricultural exports to one third of the 1928 level. In concrete terms hardship meant milk at 1.9 pence a gallon; creamery butter was 40 per cent cheaper than in 1929. Pigs and cattle prices fared even worse.

On the home market the Irish Government attempted to collect the land annuities by force and imposed tariffs up to 100 per cent on imports. This raised the cost of living and the price of farm requisites.

Table 2 Irish Agricultural Exports to Britain, 1930 and 1955

	1930		*1955*		
	vol. in '000	*£'000*	*vol. in'000*	*£'000*	*Value in 1930 £'000*
Cattle head	760	13,325	486	24,661	9,371
Beef & veal cwt	25	58	921	10,667	4,053
Butter cwt	473	2,944	3.1	50	19
Other Milk Products	–	437	–	5,750	2,186
Live Sheep head	520	1,270	95	677	257
Mutton & Lamb cwt	18	101	68	903	343
Live Pigs head	329	3,394	0.9	32	8
Pig Products cwt	477	2,347	606	8,382	3,185
Poultry cwt	120	771	140	2,937	1,117
Eggs		2,435		599	228
Total Food, Drink & Tobacco		38.5m		79.3m	29.7m

Source: *The State of Agriculture* 1960–1961, Foras Taluntais

The last column shows the value of the 1955 exports in money of 1930 purchasing power. Apart from dead meat and some dairy products it is a sorry picture.

Survival was the achievement of farmers and their co-operatives. They had no profitable investment prospect, no spirit for new ventures. It was during this period that Ireland suffered the most rapid loss of butter markets to New Zealand and Denmark.

World War 2

Judged purely by economic considerations, neutrality was a most costly policy for agriculture. World War 1 had brought prosperity to Irish agriculture. In World War 2 the British made their Ministry of Food the sole importer, able to dictate price to the Irish, who had no alternative market. Unable to buy machinery or fertilisers during the War and suffering from depressed prices since 1929 Irish agriculture was emaciated and weak. The prosperous conditions in Northern Ireland were a marked contrast.[2]

As Holmes, the New Zealand expert called upon by the Government to advise on grassland, said in 1949: '... there is no area of comparable size in the Northern Hemisphere which has such marvellous potentialities for pasture production as Eire undoubtedly has.' He saw 'hundreds of fields which are growing just as little as it is physically possible to grow under an Irish sky'.[3] Increased productivity per man employed was achieved by emigration, with minimal increase in production.

Irish Postwar Policy

Agriculture was agreed to be of crucial importance. Dr Whitaker, Secretary of the Department of Finance, stated in the 'Grey Book' Economic Development, 1958[4]: 'The importance of agriculture in the Irish economy can be gauged from the fact that over the five years 1952–6 it accounted for an average of 31 per cent of the total national income and engaged just under 40 per cent of the total number at work.' Sean Lemass could still say, in 1961: 'It is quite clear that, without a healthy agriculture we cannot expect any permanent economic progress.' The whole economy 'hung from the cow's tail'. Everything depended on an export market.

The Report on Post Emergency Agricultural Policy presumed that: 'the United Kingdom policy will be guided by ... an enlightened regard for her own national interest. While we do not anticipate any specially favourable treatment we are of the opinion that considerations of mutual self-interest forbid the imposition of unusual disabilities and call for the removal of those which exist.' They foresaw that the UK would be the only free market for produce. If it too should become protectionist, Ireland would have to reduce production. Marketing was dismissed in two pages of the Majority Report, with no mention of diplomatic activity, marketing boards, market research or quality controls.[5]

In fact Britain had already become protectionist. The Deficiency Payment System guaranteed prices to the British farmers while allowing imports at the lowest 'world prices'. Each February prices were agreed with the NFU. The food market was opened to imports from every source, giving the lowest 'world price' to the British consumer. Compensated by direct subsidy – the 'deficiency' between market price and that agreed with the NFU – the British farmer could not be under sold on his home market. Britain's open market dominated the world price. When British production increased the market was glutted and prices paid for imports fell. The reduced price of food more than compensated the taxpayers.[6]

The Programme for Economic Expansion projected an increase in agricultural production of about 4 per cent a year , all for export. The maximum estimate of expansion in British food consumption was 12 per cent over 10 years. The British Reorganisation and Marketing Policy (1966) projected increased self-sufficiency with production of beef up 35 per cent, pigs 33 per cent and poultry 25 per cent. The prospect in West Europe was little better. Consumption there might increase 13 per cent in the five years to 1960.[7] Even there self-sufficiency was increasing.

The painful lesson had yet to be learned that, for exports, every match was an away match played by the home team's rules. The industrial countries held their internal prices by restricting imports, or by direct subsidy. In the extreme case Britain, which formerly imported 40 per cent of Irish egg production, had an occasional surplus for export.

The proportion of world production sold on the 'World Market' is small (for example 3.5 per cent of meat production; 7.5 per cent of butter; 12.5 per cent of cereal in 1977). The exporting countries hold their home price up and export any surplus rest for what it will make in this narrow market. Prices fluctuated wildly in response to dumping of chance surpluses. In 1958 world milk production fell 1 per cent. Butter, the residual use for surplus milk, trebled in 'world market' price. From a level barely covering cost of manufacture, £100 a ton, it rose above the price in the protected Irish market.

Increasing production without a market was an unpleasant form of economic suicide. In 1949 Pa Quinlan, President of Macra, wrote an article in the *Irish Times* entitled 'Will Increased Production Lead to Bankruptcy?' Successive Anglo-Irish trade agreements had the objective of freeing trade with Britain by reducing tariffs.[8] However, direct producer subsidies, the new method of protection, were not covered by the agreement.

While observing the treaties Britain, by encouraging cheap food imports, had made Ireland's Trade Agreements almost useless. Britain was free to give direct payments to her home farmers. A quota could stop Irish exports when market regulation was required. There was no guarantee of access to the British market. Tariffs were removed, but when Britain had balance of payments trouble, as in 1964 a 'levy' of 15 per cent was imposed on imports. Levies had not been mentioned in the Agreements, and so were not covered. There was no court of appeal.

Only store cattle, when held on British farms for a qualifying period, were eligible for British subsidy. The arrangement gave Britain a supply of disease free stock, available nowhere else in Europe. All the employment and added value of the processing trades became located in Britain.

Table 3 Direct and Off-Farm Employment per 10,000 Cattle

	Number employed
Live Animals	2
Carcass	22
Boneless Beef	47
Consumer Cuts	87
Cooked Meats Prepared Meals	200?

Source: '*A Strategy for the Development of the Irish Beef Industry*', Irish Meat Exporters Association, 1981

The Irish farmer retained what was, then as now, the least intensive and least profitable form of farming. Even in the cattle trade Britain became a larger exporter to the Continent than Ireland, with subsidies up to £27 a head, in 1956. Britain limited the import of higher value products, such as butter and bacon, so that Irish food processing was curtailed. Livestock exports rose from 26 per cent of Irish visible exports in 1929 to 42 per cent in 1956. In 1955 the Republic was dependent on live animal exports of £40 million, being a net importer of 'other food and drink' to the tune of about £4 millions. Food processing was undeveloped.

We were becoming dangerously dependent on one low priced product sold to one market – a 'banana' (perhaps better a 'bullock') republic.

Marketing was haphazard. Forty bacon factories competed with each other, without any common standard of 'Irish'. The home market was given priority, with the factories unloading their chance surpluses on the bulk market abroad at low prices. Creameries produced only bulk butter because diversification was discouraged. The guaranteed butter price gave no incentives to initiative or quality. Livestock were exported on the hoof, giving no secondary employment. There could be no grading or incentive to eradication of pests such as warble fly or liver fluke. There were no auction marts; consolidation of creameries was discouraged.

Industrialisation

Ireland continued to use tariffs as the main protection for industry. The British paid two-thirds of the full rate, with a minimum preference of 10 per cent. Any concession given to other countries were automatically shared by Britain. The higher the protective tariff the greater the value of this concession. Britain had a near monopoly of the Irish market for industrial goods such as cars. The degree to which the market was distorted was shown in the decline in sales of British cars when the market was freed.

Ireland, with about two per cent of Europe's population, became Britain's largest European market.

Irish industrial policy was also foundering. Irish industries were geared to the protected home market, which was close to saturation and shrinking due to emigration. There seemed no prospect that these industries could develop under free trade. Foreign firms, which might have exported, were excluded by the Control of Manufacturers Act. The startling growth of industry in Ireland when industrial policy was based on exports and entry to the European market, has been a welcome relief.

A British assessment of the value of the Agreement to Irish agriculture was three per cent (about £4 million). The concession on live cattle was worth a further £5 a head (£2 million) annually.[9]

Books were written about the 'Vanishing Irish'.[10] An *Irish Times* editorial comment was 'If the present trends (of emigration) continue unchecked Ireland will die – not in the remote future but quite soon.'

CONCLUSION

Irish agriculture was a backward, high cost enterprise, dependent on a shrinking home market. National policies had deepened the world wide crisis of farming. Continuation of existing policies offered no prospect of recovery.

3

The Second Spring

The first growth of agriculture in the time of Plunkett promised prosperity. The energy of the co-operative founders, aided by the progressive agricultural views of the *Irish Homestead*, which played the role of the *Irish Farmers' Journal* in later days, set Ireland on the way to modern farming.

Frosted by disunity among farmers and politicians; hindered by unwise breeding and marketing policies, the first Spring proved false. The seedling growth had not rooted among the working farmers, who had no opportunity for education beyond the age of 14.

Education is a prerequisite for modern farming and for farm organisation. The Penal Laws forbade even primary schooling for Catholics. In the remote parts of the country 'hedge schools' continued and spread as the laws were relaxed between 1782 and Catholic Emancipation in 1829. Though the population was at a low subsistence level they paid for schooling. Travellers describe bare-foot children sitting on rocks because they had no desks. The curriculum in most schools covered mathematics, Latin, sometimes Greek, occasionally Hebrew – in effect the matriculation for the 20 Irish Colleges in Continental universities. Evidence to a Royal Commission in 1835 was that the Ordinance Survey could hire competent surveying assistants at agricultural wages. Richard Edgeworth wrote in 1801: 'I am certain that ... it will be found that not only the common but the higher parts of arithmetic are better understood and more exactly practised by boys without shoes and stockings than by young men going home on horseback or in coaches to enjoy their Christmas idleness.'[1] Official returns in 1824 showed about 7,600 Catholic pay schools, with daily attendance some half million. Present National schools number 3,400 with a half million students.[2]

The National schools, introduced in 1832, gradually took over. They gave free and compulsory primary education 40 years ahead of England. A main function, under both governments was to instil 'political correctness'. They gave uniform basic education.

Secondary education was slow to develop. The vocational schools began in the 1930s. In 1938 2,500 sat the Leaving Certificate from the secondary schools, compared with 70,000 in 1994. In the 1950s some creamery areas had no farmer with education beyond the age of 14.

Macra na Feirme

Growth came again in the Second Spring of the 1950s.

Education transformed the farmers. The crucial change for rural areas came with the vocational schools in 1930. These schools, with the enthusiasm of Chief Educational Officers, Rural Science Teachers and Agricultural Advisers, stirred an interest in education that continued beyond the classroom. From them came Young Farmers' Clubs which played the role of the Folkhighschools in Denmark.

Spontaneous growth of young farmers' clubs occurred in many areas about the same time. Among the very first clubs was that in Athy. After a lecture on 'Soils and Manure' by Steve Cullinan (CEO) and Bill Brickly in February 1944, a resolution was proposed that a Young Farmers' Club be formed. The motion was defeated. They met again a fortnight later to establish a club for all farmers. Dr Juan Greene was elected president and his father trustee. The national organisation headquarters in Athy was officially opened by Sean T. O'Kelly, President of Ireland, in 1948. Sean Healy, later General Secretary of Macra and then of IFA, entered the staff at that time.

The impact of these clubs in training leaders can hardly be exaggerated. They taught the skill of organising and identified the people who could get the work done. Debating contests made farmers articulate and taught procedures of debate – as well as spreading farming knowledge. In the early 1950s it was easy to gather a crowd of a couple of hundred for a lecture on serious topics. In villages such as Croom the lecture might go on past midnight – and sometimes a few of the audience would follow the Barrington lecturer to listen and question again the next night at another venue. There was no television then.

Among the popular subjects were co-operation and foreign farm organisations. Lecturers came from the Boerenbund of Belgium, the Netherlands, Britain, Northern Ireland, but especially from Denmark, and from the Council of Europe. Exchange visits to the '4H' clubs (rural youth clubs; 4H stood for Head, Heart, Hands and Health) in the USA and continental countries drove home the lesson. The Carnegie Trust gave a grant to employ a Secretary, Kevin Bellew, and the Duke of Leinster gave premises for a yearly rent of six pence.

It is common to criticise Irish farmers for failure to develop farming and its organisation as the Danes had done. More careful reading of history shows the Irish in a better light. Social change comes by generations and each generation has its problems and its contribution to a solution. The parallel between the Danish and the Irish time-span in farm organisation is surprisingly close. Danish farmers had been reduced almost to serfdom by about 1750. Land reform took place in 1788. It took more than a generation from the gaining of farm ownership rights to the start of the educational movement; farm co-operative organisation was the work of the following generation. Irish development took the same course.

Table 4 Tempo of Development in Denmark and Ireland

	Denmark	*Ireland*	*Years' Gap*
Freedom from Landlords	1788	1880	92
Separation from Norway/England	1814	1922	108
Folkhighschool/Macra na Feirme	1844	1944	100
Farm Organisations	1870s	1955	90

A crucial difference lay in the national experience of war. To the Danes war, even in defeat, was a unifying shock. The Irish Civil War divided the people for a long generation, overriding any constructive policies on economic or social issues. Neutrality in World War 2 was a negative experience inhibiting any national initiative.

Macra's Relations with the Department

Macra's first contacts with the Department were through the technical officers, such as Dr Harry Spain, who were fully supportive. In March 1951, under the chairmanship of John Mooney, Macra began to be in contact with the bureaucracy controlling the Departments. One aspect was the demarcation dispute between Agriculture and Education. Rural schools could not teach 'Agriculture', only 'Rural Science'. Agricultural Advisers could not teach in the schools. The leaders felt the need for a new element to stimulate progressive farming through the clubs and to put new spirit into the Department.

A relatively minor incident was influential in forming the attitude of the Macra leaders. Rabbits had become a major scourge of farmers. Seven rabbits eat as much as a sheep and they grazed the fields bare – often for ten metres out from the hedges. Australian scientists developed the myxomatosis infection to clear the pest. The disease was taken up and spread in Britain and elsewhere as a boon to farmers.

Instead of using the new method of clearing vermin, Tom Walsh, Minister for Agriculture, listened to the conservationists and gun clubs. Rabbits were to be preserved. 'Myxo' was declared a notifiable disease in Ireland. To the Macra men the lesson was clear: Government placed a low priority on the farmers' interest and could not be trusted to act on their behalf.

John Mooney and others decided, during the Spring Show of 1954, to introduce the virus from England. One of the group flew over and obtained five swabs to inoculate Irish rabbits. Farmers spread the disease throughout the country.

Irish Farmers' Journal

Once the clubs became numerous an official Macra journal was needed to keep the members of the movement in contact. It is futile to have a message, however good, if there is no medium to convey it. Steve Cullinan, a founder of Macra,

4 Tom Walsh, Minister for Agriculture, made rabbits a protected species.

started the *Irish Farmers' Journal* in July 1948 with funds provided personally by John Mooney, chairman of Macra, John Greene and Patrick Keogh. Macra contributed a subsidy of £30. Its early history was chequered. It became a fortnightly, with little impact on technical issues.

A group was formed. Edmond Sheehy, Professor of Animal Nutrition, UCD, and a trustee of Macra, with John Litton, later chairman of the Agricultural Institute, were the senior figure-heads. Two graduates of 1947: Paddy O'Keeffe, editor, then producing spectacular grassland innovations at Portrane Hospital farm, and Micheal Dillon for market reports, were recruited. These men invested £100 each and John Mooney, putting up the balance, became life chairman with a 95 per cent share. They bought the *Journal* with its circulation of 2,500 a fortnight (now 75,000 a week) from the proprietor. The title 'Official Organ of Macra na Feirme' was retained and £500 a year was paid to Macra. Sean Healy represented Macra on the Board.

Success was not immediate. The promoters faced losses of £12,000. The position of the editors was even worse. The Department at that time was almost the sole employer of graduates in agriculture. No newspaper can survive on sales. When Jack Carmody took over the sale of advertising space in January 1952, connecting progressive farming with advertising and agribusiness, the venture became viable. Independence had its price when petrol companies or the Department withdrew their advertising, or when farmers burned the *Journal* in Tullamore on the march to Dublin (1965).

The *Journal* became the leading farm paper, read on the farm, in business, in the Department and by foreigners as the best information on Irish agriculture. It was first in promoting any new idea from the eradiction of warble fly in Donegal to the spread of myxomatosis in rabbits. Its ridicule of many policies by 'Stone the Crows' or 'Ploughman' (John Monaghan) was powerful in influencing policy.

Fearing that the paper would be taken over by a purely commercial interest – as has happened with both national and local papers – or that the NFA might split the readership by starting its own journal, the owners transformed the company into a perpetual trust to which they transferred their shares without compensation (1965). John Mooney was chairman (succeeded by Paddy O'Keeffe in 1993) and NFA appointed four of the eight trustees. Profits are used for the promotion of Irish agriculture.

Without a journal the developments of the 1960s and 1970s would have been impossible.

Irish Creamery Milk Suppliers' Association

In November 1949 Limerick County Executive of Macra met at Croom. Sean Healy attended from headquarters. The farmers present wanted to discuss milk price, but Healy pointed out to them that economic matters were outside the scope of Macra, which had only social and educational functions. He suggested that they could discuss what they liked after the meeting. When the Macra meeting concluded, those present formed themselves into the Limerick Creamery Milk Producers' Association, with Liam Barry as Secretary. This evolved into the Irish Creamery Milk Suppliers' Association.

Killeshandra, county Cavan, also claims the foundation of the association. On the initiative of Charles Fletcher, the co-operative called all creamery societies to discuss James Dillon's proposed reduction in milk price. At this meeting, held in Nenagh on 24 May 1950, the national association was founded. Charles Fletcher was appointed national chairman. (Aged 60, he was a member of the Killeshandra Committee, a progressive farmer with a Jersey herd of 30. He had also a brush factory and sawmill.)

The Association was militant, concentrating on forceful presentation of the farmers' case, supported by strike action, rather than technical or market development, which it left to the IAOS.

James Dillon, Minister for Agriculture, was unintentionally the chief organiser of ICMSA when he proposed a reduction in milk price. Membership grew rapidly throughout the dairying areas. The creameries gave a natural base for organisation. Liam Barry's brother, Captain Dave, as Secretary of the Creamery Managers' Association, was a strong ally. In 1952 ICMSA could claim 460 branches. Fees of a shilling a cow were deducted from the creamery cheques. Investment in Rathkele meat factory and a box factory generated important funds.

BEGINNING OF THE NFA

As the 'young farmers' in Macra aged, their interests changed. They married and had less concern for dances. They inherited farms and became more practical. They wanted a say in the agricultural policy which governed their livelihood. Like all other observers of farm affairs they saw that the multiplication of competing organisations left the farmers with no effective representation at all. Macra would do more harm than good by adding another association where there were already too many.

Accordingly, Macra members decided the best line of development was to merge the existing groups into one union, more powerful because it would speak with a single for all farmers. An invitation was issued to all the major bodies.

Thurles Meetings

On 27 October 1953 a meeting was called by Macra in Thurles to discuss the formation of a National Farmers' Union. ICMSA, BGA and Macra attended and IAOS and the Liquid Milk Producers came to later meetings. Sean Healy, general secretary of Macra, was secretary throughout.

Fairly regular meetings were held until, on 9 August 1954, it appeared that no further progress could be made:

1) Macra and ICMSA insisted that no member of the Oireachtas should hold office in the farmers' organisation; BGA, with its chairman Martin Corry TD, could not agree.

2) The ICMSA could not agree that any individual commodity association should be bound by the decision of the federation; Macra and BGA, thinking this so cumbersome as to prevent any effective action, wanted majority rule.

Chairmanship, held in rotation between the organisations, rested with John O'Dwyer of ICMSA (father of Tom O'Dwyer, a later president of ICMSA) at this apparently final meeting. In spite of failure to reach any agreement, he instructed Sean Healy to call for one more effort on 30 August 1954. According to John O'Dwyer, the only outstanding obstacle to unity at that stage was the BGA's position on the holding of office by members of the Oireachtas.[3]

When the organisations failed to shift their positions at the August meeting, Dr Juan Greene, representing the BGA, proposed a resolution that 'the people present in the room should leave and return in 30 minutes, not as representatives of their organisations, but as individuals, to discuss the basis and means of forming an all embracing organisation'.

He then addressed them: 'If the farmers present act as a nucleus on returning to their own organisations and their own areas, and get farmers generally actively interested in a National Organisation, the outcome would be such as to enable the organisation to do what farmers should have been doing for a long time past

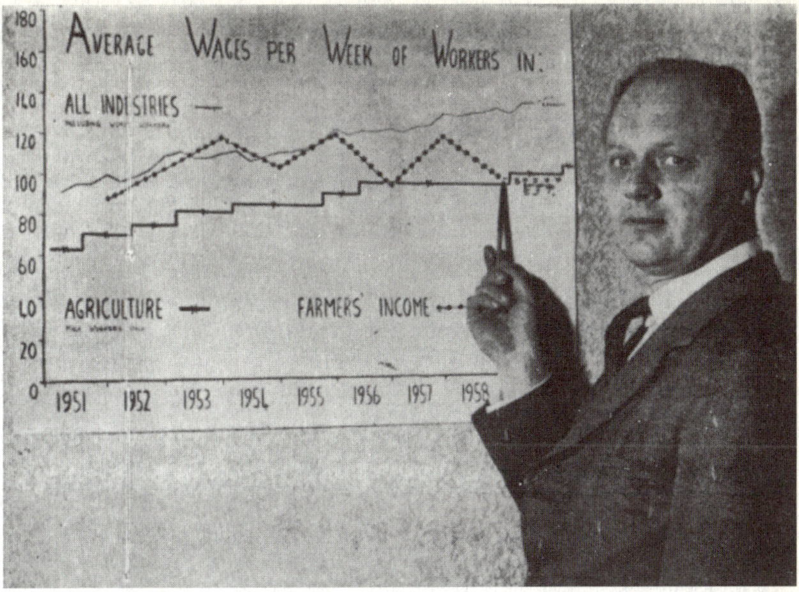

5 Dr Juan Greene

... Farmers should have their own head office and in addition all existing organisations could carry on their own good work, and, as a matter of fact, if an overall organisation was in a position to purchase suitable offices, then offices within that building could be sub-let to existing organisations.'

Harry Meredith, chairman of Macra, was then asked to describe the work of the Macra sub-committees engaged in research and the agricultural policy they were developing.

Agreement proved possible once those present discarded their organisational stand and got to work as individual farmers.

After this meeting, so nearly abortive, Macra pressed on actively with preparations for launching a union in the coming January.

At a subsequent meeting with the BGA and the ICMSA Juan Greene sketched (see p. 51) his proposed organisation on a paper table napkin during lunch in the Leinster Arms Hotel, Athy. It shows that his thought had already developed.

Preparation of Farmers

Macra had educated the younger farmers. An increasing number of them were in contact with new technology and had travelled to countries where they had seen better livestock, better marketing, better living. Their discontent with Irish conditions had the edge of desperation.

The network of Macra clubs brought these farmers together so that experienced leaders emerged to form a pressure group for agricultural reform.

Organisation was not an end in itself but a means to improve farm living. While some increase in income could be won by agitation, politicking or by a milk strike, the redistribution of income won by such means could not solve the underlying faults of low production and bad marketing. There was no prospect that the Department of Agriculture would initiate fundamental change. It would defend 'its territory'.

It was up to the farmers to put constructive proposals, since no one else would. A consistent policy had to be matured. To this end they hired Louis Smith, an economist then working with the co-operatives in IAOS, in June 1954. The appointment was part-time, providing material for lectures in University College Dublin. A range of commodity sub-committees had been formed in Macra during Autumn 1953. In addition to leading farmers, prominent businessmen and professors joined the committees. Frustrated by the Department's refusal to take advice or use research, they were eager to give long hours of free service. Individuals and groups from Macra visited and picked the brains of organised farmers in at least 12 countries. The plans for NFA relied heavily on what was seen and the advice received. The policy committees and the launch of the farmers' association, for example, almost duplicated the action of the Swedish farmers in 1930. The combination of practical farming with academic knowledge and business experience gave these committees a clear superiority over the Department.

Obtaining for farmers an effective influence in breeding policies and research; farmer control of marketing through co-operatives and marketing boards were essential if progress was to be made.

The future organisation would rest on three distinct pillars. Two pillars already existed: Macra, to train and develop people; the IAOS and the co-operative movement provided an existing base from which processing and marketing could be built. The third pillar, a farmers' union, had become essential to influence the growing of government intervention and balance the power of trade unions and large business enterprises.

The failure of the Thurles meetings during 1954 clarified Macra thinking. If existing farm organisations could not even agree on the form of a federation, on the need for which they were at one, they would never present a united front on controversial issues. Decisions, if any, would be reached too slowly and have little force. More probably the members of the federation, representing different commodities, would fail to agree and would quarrel among themselves. Any opponent could play one organisation off against an other. Friends in the Boerenbond Belge (Belgian Farmers' Union) warned that it would take 40 years to gain recognition.

The policy sub-committees of Macra in 1954 found that agreement on common policies was possible. Preliminary reports on beef, dairying, grain, pigs and bacon, lamb, wool, potatoes, vegetables were matured through many drafts.

Possible conflicts were ironed out in joint sessions. The pig producers and the grain growers were mutually dependent on each other. The committees worked together on a rational pricing and marketing systems for approval by the National Council. In case of disagreement between committees the President was arbiter, assisted by three members from each committee involved.

At the end of October the decision was taken to launch a farmers' union at a national meeting in the Four Provinces hall, Dublin, on 6 January 1955. (A holiday, the Feast of the Epiphany. The word Epiphany means 'showing forth' in Greek. In the Gospel the revelation was to the Three Wise Men of the East in Bethlehem; now it was to Irish farmers of the four provinces.) Within the Macra Council argument continued till the early hours of the morning of the big meeting on details and even on the fundamental question of whether to build NFA within Macra or as a separate organisation. Amazingly, they closed ranks on a final consensus and no trace of their doubts could be guessed from that time on.

The framework of organisation was in the 479 parish branches of Macra. All branches had to be visited and informed of developments by Macra officers and the three staff, Sean Healy, Tony Sherry and Peter Murphy. Peter Murphy was recruited by Macra in 1953 to visit all clubs within 18 months with educational films from the National Film Institute. The Department paid his expenses. In practice this prepared the ground for a farmers' union. There was debate in every club about the best form of association-development of Macra; a two-tier Macra; Federation or a newly launched association; there was doubt in no one's mind on the need for a single voice for farmers.

The Four Provinces

Pa Quinlan, as president of Macra, presided over a meeting of almost 3,000 farmers. Harry Meredith, chairman of Macra, delivered the key-note speech. The provincial vice-presidents – John Callinan (Connaught), James Hamilton (Ulster), Con Dempsey (Munster), John Litton (Leinster) – spoke, followed by Dr Greene, for BGA, John O'Dwyer for ICMSA, Ger Hill for Leinster Milk Producers, James Cashman (father of Donie Cashman, future president of IFA) for Cork and District Milk Producers, and Sean Healy.

Guest speakers were: Arthur Algeo, president of Ulster Farmers' Union, and H. Jameson, general secretary. Algeo described the 40 years of the UFU, with its 220 branches and 30,000 fee-paying members; William Howard Taft III, US Ambassador, Lt. General Michael-Joseph Costello, general manager of the Irish Sugar Company, and Pat Heskin, United Kingdom Milk Marketing Board, also spoke.

Harry Meredith began by pointing out that: 'Christian Social Thinking tells that it is the right and the duty of each vocation to organise itself in its own interest and to promote the welfare of its members ... If we can agree to pool our efforts and our resources we will not be dependent on the State to take over the

6 Foundation meeting of the NFA, 6 January 1955

functions that properly belong to the farmers themselves. Other groups of Irish farmers have considered their problem and have come to the same conclusion. With this in mind, a number of meetings have been held in Thurles to try and federate these bodies at top level, but all agreed that it would not work and the only way was to organise farmers at parish, county and in a union at national level.'

Meredith then set out the system of organisation and of finance proposed, together with the timetable for organisation:

A public meeting in every Macra area during January and February.

The newly formed branches to appoint branch officers and delegates to County Executives

County Executives to meet in March to appoint their officers and Commodity Committees.

In April branches elect National Council by postal ballot.

National Council to meet in May.

Macra would co-opt representatives of any organisation prepared to assist in the work.

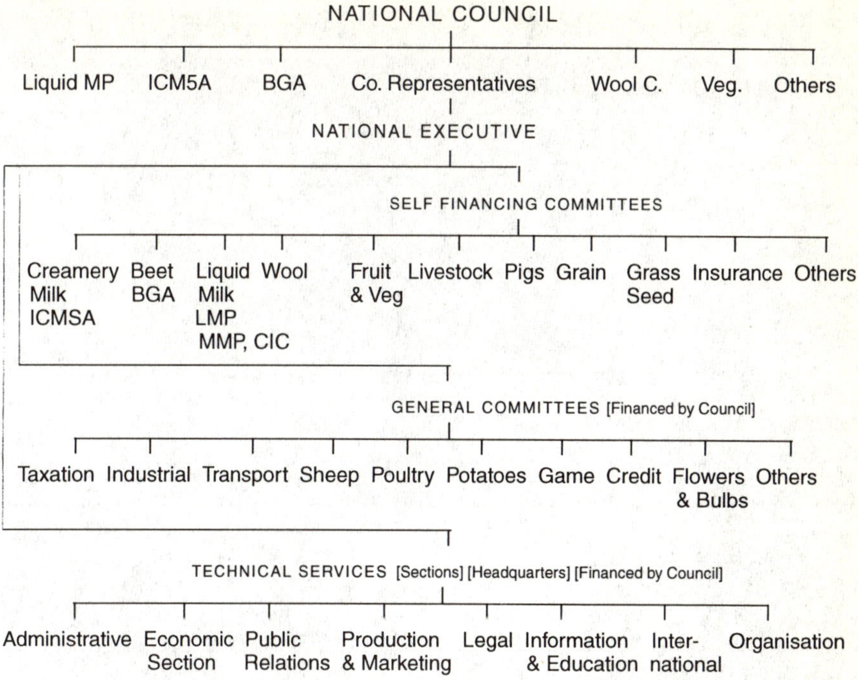

7 NFA organisation as sketched by Dr Juan Greene

Dr Greene made the formal proposal that the National Farmers' Association be formed and a provisional executive, to hold office till elections in May, was elected at the meeting.

The evening of 6 January 1956 was made the occasion for a formal dinner. Two hundred attended at the Royal Hibernian Hotel. James Dillon, as Minister for Agriculture, was the guest of honour. A famous orator, he launched into a eulogy of 'the very best Department of Agriculture in the World', how they were outstanding in innovation and recognised in their excellence by appointment to chair international meetings. He criticised the irresponsibility of the *Irish Farmers' Journal*, 'the Pravda rag of Fianna Fáil', saying that it 'pushed the views of three or four individuals, not of Macra'. He advised farmers not to read it. On a subsequent occasion, a general meeting of IAOS, Paddy Smith, of Fianna Fáil, gave the same ministerial advice.

General Costello, rose in opposition. Listing many failings of the Department and its reluctance to listen to the voice of progress, he said that the Minister's views were held by few in the country. Probably every farmer present and many nonfarmers were regular readers of the *Journal*. The General received prolonged applause. Then three cheers for the *Journal* were called, with loud response. It was a strange and unfortunate development at the birthday party.

8 First National Convention of the NFA, 31 May 1955; (l. to r.) J. Callinan, J. Cummins, S. Healy, R. Deasy, L. Smith, J. Buttimer and J. Orpen

Organising NFA

The plans were mature. The work went according to the schedule outlined by Harry Meredith to a an extraordinary degree. The first NFA head office was opened at No. 24 Earlsfort Terrace, beside University College Dublin (now the National Concert Hall). It was shared with the *Farmers' Journal* and rented from the Retail Grocery, Dairy and Allied Trades Association. RGDATA gave the hospitality of their boardroom for meetings and it was there that the first National Council meeting was held on 31 May.

The new Association could already point to successes of Macra in wool marketing; checking fertiliser price increases; orderly pig marketing. Draft policies for all major products were ready.

By Autumn 1955 a full set of County Executives were functioning, with about 350 branches affiliated. Myles Smith, vice-president for Ulster and a solicitor as well as a dairy farmer, put legal form on the Rules drafted by NFA secretary, Sean Healy. Macra paid the initial expenses of organisation and returned to its prime function of education and social organisation. It also had to be restructured, losing Sean Healy, its general secretary, two organisers and many senior members to NFA. To this day liaison and co-operating remains close on many projects.

The task of organisation was made easier by farmer anger. Virtually every farmer was a beef producer. The deepening crisis in the cattle trade, with a fall of 20 per cent in the year 1955/56 (28.8 per cent between Spring and Autumn 1956) was a major stimulant to organisation.[4] There were increases in cost of production and higher tax demands in Rates. Financial pressure is more easily understood than high ideals.

The NFA Organisation

Macra abandoned its work in the politics and economics of farming when it launched NFA. It returned to education and social activities. Drained of staff and members, major reoganisation was needed.

Staffing

NFA needed permanent staff to supplement the work of the farmers, however dedicated. Most were countrymen and had served with Macra.

Sean Healy had been with Macra since November 1948 and was general secretary from 1950. Consequently he was involved in all the preliminary work for the formation of NFA. As secretary to the Thurles meetings he knew the opinions of the other associations. He became full-time secretary to The NFA from the first meeting in 1955, becoming general secretary in November 1958. Due to ill-health he resigned as general secretary in 1976, moving to the Agricultural Relations Board with the task of improving farmers' relations with non-farming bodies such as trade unions, chambers of commerce, the Confederation of Irish Industries, consumers, till his full retirement in 1987.

In June 1954 John Litton invited Louis Smith to service the policy development committees of Macra in preparation for the new association. This could be combined with lecturing in Economics at UCD. He resigned on 6 January 1963. As chairman of the European Movement he continued to sit on the National Council. In the years of dispute with the Government there was not much room for economic argument.

John Monahan was honorary public relations officer and on the Industrial Committee. He had been brought up in Canada, where he practised as a lawyer. Legal knowledge was of value to the Committee in contests with CIE's restrictions. He served in the Canadian artillery during the Normandy invasion before inheriting his Carlow farm.

Michael Dillon advised on public relations, but his greatest and long-term contribution was in market information.

In August 1955 four organisers were appointed. Tony Sherry came from Macra, Sean Fitzmaurice, Jim Quinn (a member of the National Council for Leitrim), Brendan Sheedy (brother of Larry Sheedy, assistant editor of the *Journal*) was immediately secunded to Macra. They were recruited with the requirements of different areas in mind. Originally they were to engage in organisational matters round the counties and branches, with economists or other specialists to handle the central committees. This distribution of labour was upset by Colonel Nasser of Egypt when he closed the Suez Canal in 1956. Petrol shortage chained the organisers to their desks. Country travel was very limited. They took over from the economists as committee secretaries; Tony Sherry to Livestock; Jim Quinn to Wool, Poultry, Grass Seed. Sean FitzMaurice found his vocation in publicity and in the Grain and Dairy Committees. They learned well and fast, but gave a less academic slant than the economists.

Sean Fitzmaurice, from Wicklow, had a background in trade unions. He found his niche in journalism and publicity, but moved on to start the Grain Board as its first secretary in May 1960.

Jim Quinn, Leitrim's first representative on the NFA National Council, knew the ways of the small farmers of the West. He had remarkable courage. The job required driving to meetings, so he bought a Volkswagen. He had never driven a car, but knew about tractors. He set out for the West without further ado, staying on the road as far as Mullingar. Another time he spent a fortnight in hospital after a night on a Kerry mountain when his car stuck in a snow-drift. He became expert on matters of sheep and wool. He gained practical knowledge of the poultry trade, following the Christmas turkeys to the markets in Britain so that he could give first-hand reports on the handling defects of our exports.

Tony Sherry had already worked with Macra. His first annual report (1956) showed attendance at 163 country meetings, 64 of them inaugurals. Since meetings had usually to be in winter this occupied nearly every night. In addition he attended the National Council meetings in Dublin, canvassed for subscription fees, organised the Donegal farmers' case on contaminated seed potatoes, and other work. He specialised in livestock and the marts. In 1965 he moved to the Livestock Marts Association and used his experience in exporting calves to Italy.

Brendan Sheedy joined the NFA staff, but was seconded to Macra, helping the reorganisation, and stayed on as general secretary.

Tom Llewellyn became publicity officer after Sean FitzMaurice became manager of the Grain Board in 1960. He handled German correspondence as an added contribution.

Brendan Power became assistant secretary in 1956. He was charged with the Industrial Committee and with office management but reached his prime as marshal of the Farmers' Rights protest marches. He moved on to Farm Business Development on its foundation in 1973.

John Murphy joined the staff in 1958 and was secretary to the Industrial and Transport, the Forestry, and Game Committees.

Micheal Ross joined as an economist in 1955 when he had completed his Master's degree in UCD. A Dubliner with no farm background, he quickly learned to talk technicalities to the millers when appointed secretary to the Grain

Committee. With the chairman, John Richards-Orpen of Enniscorthy, he made a formidable team. He left the staff to join the Agricultural Institute in 1959 and, having got his doctorate, moved on to the Economic and Social Research Institute. He then studied folklore at UCD.

Joe Foyle and Monica Clune joined the staff as assistant economists after graduation from UCD, but did not stay long.

John Howley, again a graduate in commerce of UCD, was secretary to the Rates Committee. He became secretary of the FBD on its formation.

Alan Dukes became Economist on completing B.A. (with first place in economics) and his Master's degree in International Trade and European Integration. Warned that he was joining the next thing to a proscribed organisation in 1969 he saw the political opportunity. Having managed the IFA Brussels Office from its inception, he worked in the Commission itself before returning to politics as Minister for Agriculture and Leader of Fine Gael.

After the Farmers' Rights Campaign IFA advertised for an economist. Unable to decide between the merits of Con Lucey with his degree in agricultural economics and Alan Mathews, with general economics from TCD they appointed both. Dr Mathews reurned to agricultural economics in the university; Con Lucey remains chief economist of the IFA.

Richard Hourigan was appointed chief organiser in 1962, when Seamus MacCann and Dennis Knox also joined.

Dick Keenan was Accountant from 1960 till his retirement. This leaves Carmel Keogh as the only continuing staff member from the first recruitment of 1955.

NFA branches were the based on Macra clubs but the transfer of members was not automatic. The Macra staff had to reorganise the clubs, transferring members to NFA; reforming the social and educational groups with young and less experienced officers. Macra lost both members and clubs. Peter Murphy's task of visiting 500 clubs within 18 months, called for remarkable physical stamina. The £1,000 paid to Macra by NFA was a token in recognition of the services rendered.

No simple list of staff and their special work can give a true picture of the young organisation. There were no regional offices. Anyone could be called on at a moment's notice to drop the task on hands and drive a couple of hundred miles to a meeting. It was so much a part of life that Tony Sherry went to a meeting in Donegal never mentioning that his first daughter was just born, never questioning priorities. Hours of work sometimes topped 90 in a week – and there was no overtime payment. Dick Hourrigan, putting pressure on organisation in Limerick in 1964, arranged 45 meetings in one week.

The morale of the girls in the office, who worked a five and a half day week – staying on till 11 p.m. when a circular had to be dispatched – was remarkable. In 1957 22,000 letters and 42,000 circulars were sent out; 16,000 phone calls were received and 12,000 made. All that and much more, from five girls for a wage of £3 a week, equivalent to about £60 after tax now. Overtime was not regularly paid.

An example of the work load survives in the report of the Economic Section, 1962. There was one full-time economist, one part-time, one secretary/typist. The work of the Economic Section involved, among other things:

- Weekly: preparation of graphs and articles for *Irish Farmers' Journal.* Translation of French version of *Agra Europe* news letter.
- Fortnightly: preparation of statistics for Bacon Consultative Council.
- Monthly: attendance at IFAP, usually in Paris. Attendance at NFA National Council and Executive.
- Quarterly: milk statistics for Confederation Europeanne d'Agriculture.
- Half yearly: review of farm incomes.
- Annual: statistics for IFAP; cases for counties on Rates.
- Occasional: attendance at 120 meetings in Dublin and 20 county meetings.
- One TV appearance, 14 broadcasts, 12 outside lectures. This in addition to general availability.
- Researching and publishing Common Market booklet.

Even salaries were precarious in the first years. Payment was a matter of touch and go as cheques came in from the counties. More than once Sandie Gallie, treasurer of North Kildare, found money for the week's wages in emergencies or Killeshandra Co-op made an advance payment of fees. Naturally there was no overtime payment. It was normal to leave the office after lunch, travel perhaps three hours to an evening meeting at eight, talk to the officers in some hospitable house till after midnight, then start back to Dublin – and back to the desk next day. Even office furniture was scarce. Returning to the office the first task was to find a chair.

Similar dedication of officers and paid staff is found in many young organisations. It is part of the seed capital invested. Healthy maturity brings a more even balance of effort and reward. Associated Industrial Consultants were called in to guide the Council in establishing a more ordered routine. They reported in March 1963.

The consultants found: 'The current annual revenue of approximately £20,000 is inadequate to support the organisation in any satisfactory form, consequently the staff at central office is over-committed and underpaid and is not large enough to do the work that should be done.' They recommended that salaries should increase 60 per cent to a level found in other Irish organisations. The consultant argued that similar work had to be done in each national farm organisation, regardless of country size. 'It is interesting to note that the staff of the Economic Section of NFU number approximately 20. In theory at least, the field of work to be covered by the NFA Economic Section is similar' – with one economist.

The disputes of 1965 to 1970 prevented any serious effort at restructuring Head Office. The work load increased and new staff were recruited for work with the counties and branches rather than for policy formation.

When the dust settled, a more normal routine had to be established, with something like civil service procedures. A similar transition occurred in similar organisations abroad. A staff of 60 with 12 country offices and a salary bill of £1 million cannot be run on informal goodwill. The National Council minutes show that the transition was not all smooth. The consultants had stated that no more than seven staff can directly report to one senior. With greater numbers the overloaded senior has no time for mature thought. When staff, accustomed to being a group of equals reporting to one general secretary, had to be divided into grades with defined areas of responsibility, friction was inevitable.

A second consultancy report was prepared by Reynolds and MacCarron, accountants, in November 1976. The report was not circulated to National Council.[1]

Finance

Any new organisation needs outside finance while the seed germinates. The IAOS was subsidised by Plunkett, the Government and by the English consumers' Cooperatives Union. Allied Irish Banks funded management courses for committeemen and Bank of Ireland endowed the Centre for Co-operative Studies in University College Cork. Macra was given its offices in Athy by the Duke of Leinster at a nominal rent; the Carnegie Trust gave starting funds and the Kellogg Foundation also targeted Macra as an organisation with potential as a focus for social development. The same Foundation presented the Irish Countrywomen's Association with their college, An Griannan, at Termonfeckin, the Horticultural College (1968) and paid for study tours by their chairman, Mrs Hughes, and their secretary, Doreen Smith. The development costs of NFA were carried by Macra till the launch in January 1955. From that time it had to struggle for viability. In due course the NFA, in its turn, returned money to Macra and subvented the start of other organisations with which it found common cause, like co-operative ventures or the European Movement.

The first subscription was 2.5 pence per £ Poor Law Valuation, with minimum 50 pence. It averaged the price of a Sunday paper or three boxes of matches. Trade unions had minimum subscriptions of £3.50 at that time. Members were enrolled at local meetings, but more often volunteers called from house to house. The register of local ratepayers set the target. County Wexford claimed 100 per cent membership. In 1962/3 the consultants found that, while the income of headquarters was rising to £20,000, there was no prospect of income adequate for NFA's work from the current scale of fees, even if 60 per cent of potential members joined. The main income would have to come from the 122,000 farms over £20 PLV. Small farmers' contributions could not contribute much to overheads. The old subscription rate, allowing 25 per cent of fee to the branch, could only yield £55,000 at maximum. A fee of 6 pence per £PLV was adopted, with a target for central funds of £82,000.

By 1965 the canvassing members were tiring of the annual call begging from their friends. Six organisers were recruited and the subscription was doubled. Creameries (except for about 30, mostly in Limerick) deducted NFA fees for the 75 per cent of members who were suppliers. This, it will be remembered, was a point of friction with the ICMSA.

Farm organisations are expensive compared to industrial trade unions, whose members are easily contacted at their work place, or at least on routes of public transport. Farmers are scattered all over the country, with no daily meeting place. Travel expenses in the NFA were two-thirds as large as salaries, twice the office costs. Headquarters staff were paid 1.6 pence a mile. Local officers carried their own costs and there was a strong feeling that the National Council and Executive members should get no expenses. As a compromise, members of national committees could submit expenses, if in need. Their names were then 'put in the hat' for payment by lot as funds became available. Dr Greene copied Horace Plunkett. In no need of an expense account himself, he collected his money in full and repaid it to the NFA as an anonymous subscription. He did not wish to establish a precedent which would prevent poorer farmers from taking their place in national affairs.

Social and physical difficulties were more important than financial cost to farmers called away from home to attend meetings in Dublin or at county level. In the nature of things these were most often on winter nights. Members would not come to meetings before nightfall stopped farm work. Travel was a hardship, greater before cars had heaters. In the unheated parish halls, before rural electrification, the wind from the broken windows would blow out the candles. There is now a community hall costing £1.5 million. Staff and NFA officers lived off the country. That might mean changing a baby's nappies while the local chairman and his wife hurried the milking so he could get to the meeting (at Cooldaf); it could mean dining off solid gold plates as guests of the Government of India.

Foreign travel is a special burden for an island member in an international organisation, as it is for Western counties nationally. Sean Lemass had pointed out that these contacts were necessary in the national interest and should be funded by Government (see chapter 4, Appendix). This promise was not fully implemented. Later, the agreement with the ICMSA, by which representatives of both organisations had to be represented, nearly doubled the cost to the Irish farmer. Only at the FAO general meeting in Rome did Sean Healy discover, in casual conversation with a Dutch colleague, Rinnse Zylstra, that the other

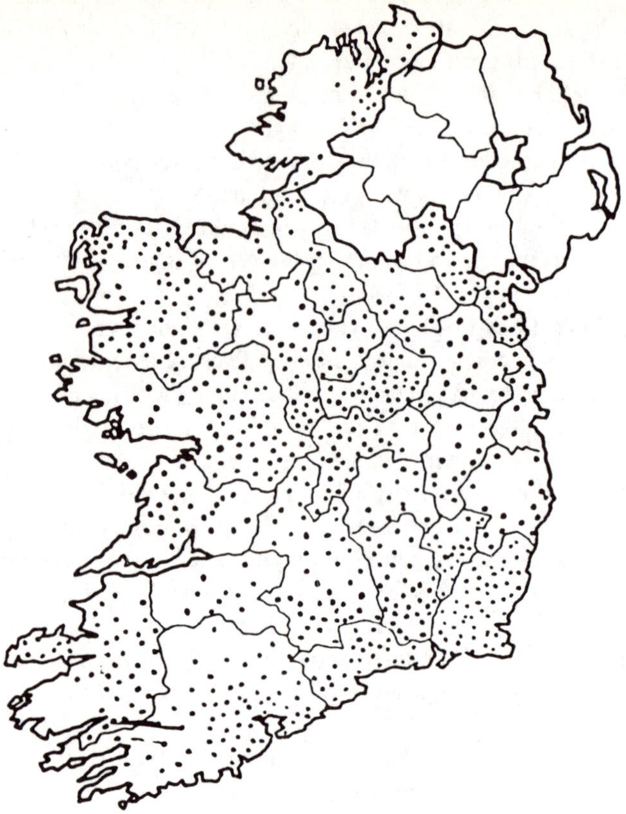

9 NFA branches throughout Ireland, 1959

countries representatives had their expenses paid in full, half by FAO and half by the national governments. In spite of Sean Lemass's agreement neither NFA nor ICMSA had been informed.

Special funds were opened for particular projects such as the purchase of NFA House and, later, of building the Irish Farm Centre, was already in view. Seeing the experience of all farm organisations a fighting fund was collected in 1965. At the general meeting of 1964 David Dolan, national treasurer, reported that the finances were sound. Fee income, 1963, was £64,000 (of which £18,000 was retained in branches). Various items brought headquarters income to £56,000, showing a surplus of £3,000. The IFA was in funds for the first time. He warned of the danger of complacency: 'A tremendous amount of work still remains to be done. Further appointments, as outlined in the Consultants'report, remain to be made. Costs under almost all headings are increasing ...'.

The dispute with the Government was on the horizon and a fighting fund was needed. The Association's bankers indicated that they would not consider increasing the over-draft to cover the cost of disputes. (See Appendix on Finance of IFA in Chapter 13.)

Premises

The first offices were at No. 24 Earlsfort Terrace, over the premises of RGDATA, who also made their boardroom available. It was a happy relationship, but as soon as office staff became employed larger premises were found down the road at NFA House, No. 27 Earlsfort Terrace. (The *IFJ* was in No. 24.)

The NFA bought NFA House for £6,000. Pressure of space moved Brendan Power and the Poultry Committee to No. 28 Adelaide Road in August 1963. When building the Irish Farm Centre NFA House was sold for £22,000 in 1973 and the lease of No. 28 for about £15,000. The IFA share in the Farm Centre cost £30,000.

Commodity Committees

The vital interests of all members were linked to the national groups through branch and county committees. Like farmers, businessmen and academics felt ignored by Government. Consequently, the National Committees could call on frustrated experts with practical experience in production, processing and marketing. Such were Professor Edmund Sheedy and Dr Brendan Senior and cattle dealers Joe Bruton and auctioneer Bill Smith who gave their knowledge and time to make NFA often better briefed than the Department.

NFA committees were the work place of the organisation. Any topic of current interest was passed to a small group for study, policy recommendation and action. Description of their work easily becomes a jumble of detail on hardly related topics. Basically the work came under three headings:

1) *Technology*. Irish agriculture lagged far behind its competitors. In Macra the practical farmers realised the deficiencies. Finding no existing structure for influencing national policy, they joined with academics and business interests who were similarly frustrated.

2) *State administration*. Regulations affecting farmers were introduced without reference to the needs of the industry. This was particularly evident in transport, industrial protection and in taxation.

3) *Internal affairs and administration of the Association.*

For the first ten years the NFA concentrated on technical progress. Frustrated in these efforts, agitation took precedence in the second half of the 1960s. In prospect of entry to the European Community the emphasis on foreign affairs increased.

BREEDING POLICY

All commodity committees shared problems in breeding.

The Departmental livestock policy, from its foundation in 1900, aimed at uniformity of stock with few selected breeds. The Livestock Breeding Act 1925 strengthened the Department's control by making it an offence 'to own, or to have in possession', a bull or boar without a licence permit. The criterion for selection was visual inspection at crossroads or other centres. This gave no indication of fodder conversion, milk yield, distribution of fat. The result was a beauty contest, judged by fixed rules with little relation to value in the market. To take a petty example, points were awarded for the type of horns. It was a couple of hundred years since the last wolf had been killed and horns were no longer needed for defence. As it was said, 'cattle went like a regiment with fixed bayonets', injuring each other in transit by road – or especially at sea. It took years of representation to have regulations changed to qualify poled cattle for show prizes.

Any advance was limited by the unscientific base of these schemes. It was only in 1958 that Macra introduced a competition to the RDS in which judging was first done on live cattle and then on the carcass value of those same animals. Progeny testing, introduced by Danish farmers about 1900, was introduced to Ireland half a century later in belated response to the farmers' demands. Legal import of improved stock from the Continent was impossible because there was no quarantine station. Farmers acquiring Landrace pigs were prosecuted.

Lest the reader feel that these comments on breeding policy are intemperate, a more scientific statement may be quoted: 'The need for documentation of this work (of breeding) led to the establishment of herd books and, with the idea of the herd book developed the idea of 'pure Breeds', whose purity was protected by rules for registration. The herd books were the only documentary evidence of an animal's worth, and over the years excessive reliance came to be placed on the pedigree registration. The development of agricultural shows in the last century and the prominence that was given to show winnings in the breeding world led away from the evaluation of animals on their productive merit towards a greater emphasis on points of type, whose relationship to productive capacity was often extremely tenuous. Thus by the beginning of the twentieth century, the methods employed in many breeds for the evaluation and improvement of stock had strayed a long way from those used so successfully by the pioneer breeders of two centuries ago'.[1]

One of the first acts of the Department was to import pure beef Shorthorn bulls from Scotland. The Shorthorn breed might have proved suitable if an effective improvement policy had been followed. In basing policy on the 'Dual Purpose Shorthorn' they did not know that the difficulty of improving two characteristics in a breed is not twice but four times as great as when concentrating on one. Besides, the steps taken to implement it were misguided. Some experts believed that milk yield was sex linked, therefore the bull could not affect its daughters' performance.[2]

LIVESTOCK COMMITTEE

Michael Gibbons, who repeatedly features in the early NFA records, was chairman of the Macra Livestock Committee (1954). The Committee found a depressing scene. Cattle formed 40 per cent of Ireland's exports, but the system was chaotic and costly. Cow numbers had remained static for 100 years. Eradication of TB, contagious abortion, warble fly, control of calf mortality were essential but neglected. Scientific research in breeding was absent. Marketing within the country in fairs was costly; there was no plan for the orderly development of the export market for Irish meat. Animals were exported live, giving a minimum of employment and value added in Ireland. Grassland was unimproved and silage a recent introduction. A depressing picture for the country's major export industry.

Breeding

Breeding policy had accidentally favoured the beef industry. Selection on visible conformation, effectively ignoring milking quality, produced a beast fairly suitable for the British market. Better stock, specially suited to lean beef production was available on the Continent. During the NFA Committee's visit to France in 1962 contact was made with the breeders of Continental stock, who offered foundation stock of Charolais and other types suited to the lean meat demanded in that market. The Department refused to allow import. While some stock filtered in through Britain, the availability of the best stock was set back eight years.

The Disease Sub-Committee

The first task was to establish a priority of need in disease eradication. A survey was made of the action taken in other countries. Then the veterinary members prepared papers on the main diseases.[3]

Brucellosis (undulant fever in humans) had been eradicated in Norway in 1944; Sweden and Finland, 1956; Denmark,1959; and virtually eradicated in Germany, France, Italy, USA and Canada. Britain and Northern Ireland were commencing. It was costly to farmers and dangerous to humans.

Tuberculosis eradication was urgent. Clearance of the disease in Britain could bring a ban on imports from Ireland, destroying the whole live cattle trade. The NFA Animal Health sub-committee, which included vets and dealers as well as farmers, met Sean Moylan, Minister for Agriculture, and the Department officials to propose a TB eradication scheme which had begun in Donegal in 1954. The sub-committee favoured attacking the worst areas, sources of infection, first. The best areas, such as North Donegal with less than one per cent reactors, could then be cleared without danger of reinfection from bought-in stock. The Department was in favour of starting with attested areas to make them 100 per cent clear. Moylan listened and gave judgement for the sub-committee. After his death the Department scheme was implemented. The country was declared attested in 1965 but, £1 billion pounds expenditure later, the problem remains.

Warble fly was widespread – 80 per cent of untreated two-year-old cattle averaged 20 warbles. This reduced the value of hides by 50 pence a beast and cut milk yields 10 to 25 per cent The total loss was estimated at £764,000 a year. The Warble-Fly Order of 1936 was unenforced. Buyers of live cattle, unable to tell which beasts were infected, reduced the price of all by the average loss. Thus the loss was suffered by all farmers without any incentive to improvement. It was futile for an individual farmer to treat his cattle when the insects could fly in over the hedge.

Donegal County Executive instituted a programme of warble eradication in 1955 and eradication in the whole country was within sight in 1967, when the Department scheme was suspended (in spite of NFA representations). The pest spread again until, with a renewed effort the final eradication was achieved in the mid 1980s.

Liver fluke affected 95 per cent of all cattle in Ireland according to the OEEC estimates. Depending on rainfall, 50 per cent to 95 per cent of sheep livers were condemned. Livers of 75 per cent of cows and 35 per cent of beef cattle were condemned. The net loss was estimated at £1.09 per beast.[4]

Irish live exports depended on a disease-free reputation. Reports were received in the 1950s that British farmers were unwilling to risk purchase of Irish stock for fear of contamination. For example, sheep scab had disappeared from Britain in 1940; brucellosis could bring serious loss to the British farmer.

The only Irish veterinary clinic was in Castleknock, county Dublin. The vets advised that one was needed every 50 miles.

The migration of cattle between farms and around the country; the contacts of cattle from different herds at fairs and marts has made the final eradication of disease almost impossible. The farmer who treated his yearlings gained nothing in the price; the final purchaser had no way of knowing the condition of the beast's liver. All cattle, clean or diseased, are cut in price. There has been an opportunity for short term gain in smuggling; in moving cattle from diseased herds; in illegal hormone treatment with growth promoters. Efforts to focus farmer-opinion against the culprits so that heavy fines might be imposed have not been successful.

Cattle Expansion Scheme

Stock-carrying capacity was increasing due to reclamation of a million acres, better grass management and increased use of fertilisers. The herd was being reduced by the slaughter of TB reactors. NFA set a target of 50 per cent increase in exports in seven years to 1970, requiring an increase of 350,000 to 400,000 cows.

Farm profits could not generate the capital necessary for increased stocking. Farmers could not afford to breed or buy replacements, especially after the bad harvest of 1958. Farmers were not known to the banks.

The problem was largely solved by the Heifer Loan Scheme (described later under the Capital Committee) to which the cattle marts, the Veterinary Association and the Department gave full support. It seems the time was ripe. In 1958 35,000 heifer loans were given, in 1959 £2.5 million. Heifers increased 17 per cent in 1959.

The ICMSA denounced the NFA TB-free heifer loans as 'humbug'; inflating the price of replacements for the creamery supplier without doing anything to eradicate TB.[5] The NFA scheme was 'degrading' and 'gombeenism': 'The self-styled National Farmers' Association plan does not provide for any measure of financial protection for the dairy farmer against loss (in fact insurance was provided shortly afterwards); on the contrary it proposes to make him a debtor to the banks at the current rate of interest for a period of two years. What an absurd proposition.'[6] ICMSA and NFA proposals that full compensation be paid for reactor cattle and for a bounty system, had been rejected by the Government.

Marketing

Dependence on live cattle exports was not a free and happy choice of Irish farmers. Intensive production of of perishable, highly priced crops are naturally developed near the growing industrial cities. More distant land was left to grazing. Industrial countries, aiming at self-sufficiency in case of war, shut Ireland out of their markets, keeping their redundant farmers happy by high prices for intensive agriculture. Any imports necessary were taken in the form of live animals. Ireland became a source of live cattle for the British market. Low prices for stock could not out-weigh the effects of British subsidies on the prossessing industry.

When exported to Britain, stores were fattened for a few months to be 'naturalised', qualifying for British home market guarantees – 10 per cent of 'British home produced' meat was Irish, sold as British and commanding a premium over the same meat exported from Irish factories.

The Livestock Committee visited the NFU and Meat Marketing Board in 1955 to meet their customers. Michael Gibbons visited the Paris market in June 1955. The need for an Irish meat marketing board was stressed by all. Frequent contact with Paris and other markets was kept up thereafter. Particularly close contact was maintained with the UFU, sharing as they did the interests in disease eradication. In the market assessment meat was classified as Irish regardless of the Border. Reports were shared with the Department and meetings kept contact almost weekly. Relations were eased by the fact that James O'Mahony, secretary of the Department of Agriculture, William Hooley, the deputy secretary and

Patrick O'Higgin, Department of the Taoiseach, had been class-mates of Sean Healy in Farranfore secondary school in West Cork.

Britain could not absorb as stores the increase in exports from 850 thousand head to 1.3 million implied in the the Programme for Economic Expansion (1958). The instability of the market was shown in 1956 when imports of Argentinian beef coincided with a shortage of fodder in England, reducing the demand for Irish stores. The price fall was quickly reflected to yearlings in Connaught and calves in Limerick.

The Committee report of 1964 warned of the danger of a glut – even without the planned increase in Britain's home production. Prices would collapse unless an active marketing development was undertaken. This, they argued, could only be done by a marketing board. Sean Lemass, at first sceptical, was convinced that action was needed and a meat marketing board was promised.

Nothing was done. The worst fears of the farmers were realised when cattle numbers increased to 4.9 million in 1963 and 5.6 million in 1966. The Government relied on the Trade Agreement, but this placed no obligation on British farmers to buy the increased store cattle offered. It was this inaction which added acrimony to the disputes of 1966 when cattle prices again fell by 23 per cent.[7] The Meat Marketing Company, CBF, (not a Board) was established in 1967.

The Marts

The last major fair in Britain was held in 1926. Most continental countries had modernised their marketing years before that. Ireland still relied on fairs held in the streets of towns and villages on dates fixed from time immemorial. The farmer walked his cattle to the town and stood by them as the dealers strolled round or 'tanglers' latched onto him, standing by the cattle so that no other buyer would approach. There was no way for the farmer to know the current state of the market. He bargained against men who had inside knowledge, in a system that was time wasting and hard on men and cattle.

Daily publication of prices from Radio Éireann by Michael Dillon and in the *Farmers' Journal* helped close the knowledge gap.

Direct sale of cattle from farm to factory, minimising cost, would have been ideal. The farmer could be paid the true value of his product according to grade and weight, with deductions for damage from liver fluke, warbles etc. The system of payment used in Sweden was known.[8] Until the farmer who treated his cattle for parasites could get payment into his own pocket he would grudge expenditure on eradication. While cattle were moved from farm to farm and exported live such payments were not possible.

The NFA launched a campaign to establish cattle marts throughout the country. The auction system was second best to direct factory sales, but a great boon:

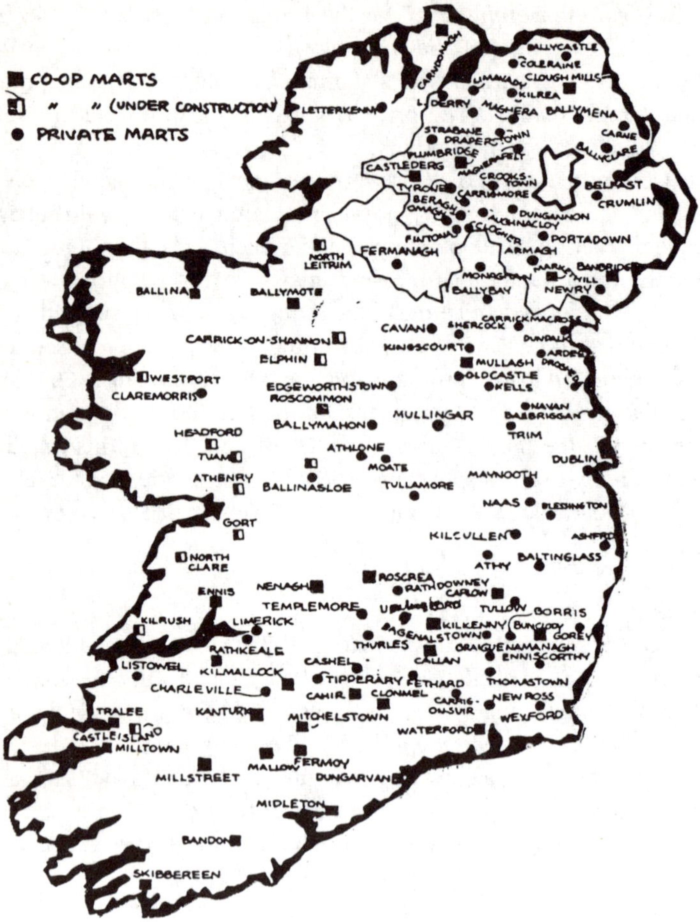

10 Distribution of livestock marts, *c*.1958

- farmers were in direct contact with the buyers, cutting out middlemen.
- weighing and open bidding established a market price.
- cattle could be offered in matched groups convenient for the buyer.
- the farmer was shown the breeds and types that the market demanded.
- buyers and sellers had shelter and decent amenities; the townspeople clean streets.

- auction marts were a vital step in the eradication of TB. Disinfection of fair grounds was impossible, so the cattle sold in the old way lost their accredited status.

Not surprisingly, the marts were denounced by the Irish Livestock Exporters' and Traders' Association: 'Marts have made the successful operation of the trade unworkable. It is only through the sale of cattle in the fair and on the land that the trade can properly be carried out. Ninety per cent of the cattle traders have tried the marts and found them impractical and unworkable'.[9]

A letter from the Livestock Traders' Association threatened to boycott any Dublin salesman handling cattle which had passed through a mart: 'It was a job that was done for a livelihood. We are going to fight to retain that livelihood'.[10] The NFA pointed out that the same traders sold through marts in England.

Still, the spread of marts was phenomenal. The NFA branches took the initiative in promoting local marts, especially in the West, where no dairy co-operatives gave a base for expasion. Then the IAOS registered the society and gave technical advice. Waterford and Kilkenny marts were working in 1957, with Cork, Tipperary, Limerick being organised. By 1960 there were 30 marts and half a dozen in the pipeline. Ennis Mart was the first west of Shannon (10 January 1959). Its success gave the Connaught farmers the confidence necessary for co-operative development in the province. The Taoiseach, Sean Lemass, and Sean Moylan, as Minister for Agriculture, officiated at many opening ceremonies.

The cost was about £15,000 a mart. If 1,000 members paid £15 in shares and the gain in value was, as claimed, £5–10 a head, even the smallest farmer got his money back within a year. The banks agreed to a loan of £10,000 to cover set-up costs.

Marts were located where transport and supplies seemed best. Years later, drinks on the house could be called when one of the original organisers called in the nearby pub. But not everyone was happy. Shopkeepers disliked marts. The farmer was paid by cheque in the post, so there was less loose cash to spend in the town on the fair day. With the money in the bank farmers took their time to shop round for the cheapest prices and more goods were bought at the co-op store. Marts needed to be at least 20 miles apart to collect enough stock to attract a crowd of competing buyers. In towns with no mart both pubs and shopkeepers lost customers.

Every town wanted a mart to retain its trade. Merchants would subscribe capital or persuade the banks to finance local marts. Instead of a mart every 20 miles, they sprang up all over. Ballinasloe had two. The development wasted capital, raising costs per head sold. The NFA, expressing concern at the wasteful development asked the banks to refuse credit for redundant marts.[11] Banks saw the force of the argument, but found it hard to refuse applications for loans supported by valued and solvent customers. In January 1959, after a meeting of representatives of NFA, IAOS, Banks and the Department, an umbrella organisation, Co-operative Livestock Markets, was formed to co-ordinate development. Co-operative and private marts worked together in the marts Permanent Committee (April 1959). Though farmer owned co-operative marts were the preferred choice of the NFA, all were interested to co-ordinate dates of auctions; advertisement of sales and promotion of attendance by dealers; preparation of a black-list of insolvent buyers and other matters. This was the task of Tony Sherry of NFA staff.

Banks agreed to give preference to new marts sponsored by IAOS and NFA. Understandably proposals for small, uneconomic marts gave rise to further consultation with banks.

There is a limit to what can be done by a voluntary body. Repeated requests for Government regulation – preferably through a producer body, otherwise by Government – were ignored. Eventually, Government did the right thing, in the wrong way, by the Marts Act, 1967. This was aimed as a blow in the dispute with the NFA. The Act gave such arbitrary power to the Minister for Agriculture that the NFA appealed, unsuccessfully, to President de Valera to refuse signature. Appeal was made to the Supreme Court to have it ruled unconstitutional. This was the first use of the Constitution to upset legislation. It was not an outright victory, but NFA were awarded costs and the Government advised to amend the law.

Meat Exports

NFA relations with the Fresh Meat Exporters' Association were immediately friendly. The Association was formed in 1941 to relate with the British Ministry of Food, the sole importers to Britain prior to 1954. There was a common interest in improving the primitive conditions of a newly developing market. There was no refrigerated transport; sides of beef cut 'on the chine' were stacked in the ship's hold; handled on to the floor of an ordinary railway wagon for carriage to the British city markets. Shippers refused to provide hanging facilities till the traders association paid half the cost of meat hooks. Exports in insulated containers began in 1952.

The NFA immediately related to this dynamic industry. It gave industrial employment in rural areas in slaughter and in processing of hides and offals. It reduced transport cost. It reduced dependence on the store trade, which was known to be unable to absorb the planned increase in livestock production. It broke the monopoly of British buyers. The Association never sought a full dependence on beef rather than live export.

The marts and meat factories amply repaid the NFA for the effort in development by providing a reliable way of collecting funds from members.

DAIRY FACT-FINDING COMMITTEE

Cattle and milk each formed 38 per cent of agricultural output. The situation reviewed by the farmers was bleak. Milch cows numbered 1.2 million when first counted in 1861. There was no significant change in all the years to 1955. Milk yields were half those in competing countries and in decline. The *Interim Report on Milk Production*, 1920, estimated the average milk yield at 420 gallons a year, but it took the price of 400 gallons to cover the maintenance cost of a milch cow, excluding labour. The official estimated yield of 445 gallons in 1953 was shown by the NFA to be exaggerated. It seems incredible now that the Report on Post Emergency Agricultural Policy (1945) should state as a prime dairying objective,'It is desirable in the national interest that a supply of butter adequate to the requirements of the home market should be forthcoming as soon as possible ... If and when any export surplus becomes available ...' (S. 237). In 1951, 1952 and 1953 Ireland was a net importer of butter by a margin of £1.8 million in 1953. Exports of eggs were 23 times, rabbits nearly six times, the value of butter exports.

The Dairy Fact-finding Committee of Macra, so called to avoid hurting the feelings of ICMSA, was of highest priority in preparation for a farmers' union in October 1954. As things turned out the Committee was constantly involved in time-consuming interorganisational negotiations. It was forced, almost at once, to go beyond fact-finding to active farmer representation to fill a vacuum in representation abroad. The committee combined the academic with the practical: creamery co-operative board members, such as Con Dempsey of Waterford, Jerome Buttimer and P.I. Meagher, later presidents of IAOS; academics, John Lyons, Professor of Dairy Science in Cork and Edmond Sheehy, Professor of Animal Nutrition, UCD.

Breeding

Dairy farmers suffered from the Departmental breeding policy. The 1920 Report found: 'We feel that we cannot but concur in the consensus of opinion that bulls being generally selected for breeding purposes, while undoubtedly improving the beef producing qualities of the Irish Dairy Shorthorn, have achieved this object at the expense of their milk yield.'

The Cow Testing Association, begun in 1909, might have drawn attention to the decline in milking quality of stock. In fact only five per cent of herds were milk tested in 1962, and the number of cows tested fell again by half between 1965 and 1970. This may have been either the cause of the problem or its effect. In spite of the recommendations of the *Interim Report on Milk Production*, 1920, and the Commission of Inquiry into the Resources and Industries of Ireland, 1922, and the example of other countries, no use was made of milk records in national breeding policy. There is little use in looking at a speedometer when stationary or sliding backwards. Farmers lacked the basic knowledge and guidance necessary for the selective breeding and progeny testing, used by the Danes since 1900.

The NFA Dairy Committee from the start advocated that development policy should concentrate on a specialised dairy breed with efficient food conversion. Like the beef producers they looked to the improved stock already available on the Continent for a leap forward by importation. Without quarantine facilities only imports of Shorthorn and Friesians from Britain were possible. The latter were bred for volume of liquid milk with low fat content, unsuited to creamery production. A quarantine station was established in Cork in 1963, allowing import of Charolais from the USA in 1963, and Charolais and Friesians from France and Netherlands in 1964.

The real breakthrough came from the development of artificial insemination by the co-operatives. Stations were started in 1947 at Ballyclough and Mitchelstown. The Department set up stations in other areas, most of which were transferred to co-operative ownership before 1965. One serious reason was the cost of bureaucracy. In Killeshandra an inseminator's car broke down. It had to be towed to Dublin, where three quotations were obtained. The spark-plugs were replaced. A car had to be rented during the operation. Trade union difficulties arose. Was a university degree necessary for a man to replace a bull? How to establish the farmer's right to 'bull his own cows'?

The Department gave strong support to the new system, though they were loath to yield their policy of promoting dual-purpose Shorthorns. An attempt was made to influence breeding policy by subsidising Shorthorn insemination at twice the rate for other breeds.

With the prospect of progress in dairying, Shorthorns declined from predominance to one per cent of dairy herds, which became 97 per cent Friesian.

As the cattle of Ireland turned from brown to black and white, milk yields took an upward trend. Output increased from 489 million gallons in 1957 to 1,245 million in 1992, though cow numbers increased only 20 per cent.

Farm Butter

Milk output figures were of political significance. The Government wanted a high figure to show the success of their breeding policies. The NFA wanted to show a low farm income, building a case for a higher milk price.

The official method of calculating milk yield was faulty. To the known figure of milk sold was added an estimate of milk fed to livestock and another guess for human consumption on the farm. The residue was presumed to make farm butter. The authorities, presuming the success of their breeding and other policies, increased their estimated yields each year. The residue unaccounted for, attributed to farm butter, rose till it reached half the figure for creamery butter in 1955. This was contrary to all knowledge. The Government fought hard to maintain their figure. The NFA pointed out that farm butter production was dependent more on the declining number of farmers' wives and their declining willingness to churn butter, so the Department figures must be wrong. The official estimate of milk used for farm butter was revised downward from 112.6 million gallons in 1957 to 29 million in 1967 and 2 million in 1980. No official explanation was given.

The reduction in estimated farm income by about £4 million strengthened the NFA case for farm income stabilisation and for a change in breeding policy.

Dairy Policy

The Fact-finding Committee was forced almost at once to intervene in dairy policy. The ICMSA was by its structure limited to protest and applications for better prices. In its early years it did not have the personnel, or the contacts, to form a general policy or to represent the dairy farmers interests outside Ireland. The Dairy Committee filled the vacuum.

The Committee's recommendations to the Advisory Committee on Marketing of Dairy Produce were accepted, leading to the establishment of Bord Bainne. Disappointingly, the recommendation that election to Bord Bainne should be by direct vote of suppliers was not accepted at that time.

NFA recommendations for action on tuberculosis eradication, more far-reaching than the legislation, were not accepted by Government. A similar rejection met the repeated requests for milk quality incentives – eventually independently introduced by the co-operatives.

Processing

As described in Chapter 1, manufacturing creameries were consolidated by Patrick Hogan, Minister for Agriculture in the 1920s. The system was still geared to horse-drawn transport, with a network of receiving stations each serving a radius of an hour's journey (under five miles). In the smaller separating stations supply was down to 90,000 gallons a year.

Little progress was made in the IAOS proposals for rationalisation of creameries till the prospect of a Common Market became imminent.

In the Hogan reorganisation all the creameries co-operative and private were bought by the Government owned Dairy Disposal Company. As its name suggests, the object of the company was to sell all creameries back to the farmers in efficient units. The company continued to handle about one third of the output. After 25 years it considered itself permanent, resisting farmers' attempts at a co-operative take-over. Con O'Leary, as a new graduate , was employed by DDC to persuade its suppliers to avoid co-operatives. As manager of Imokilly Co-operative he went on to be a practical co-operator, named Co-operator of the Year in 1996. The Committee on Marketing found that the DDC paid as much as 2.5 pence a gallon less for milk than neighbouring co-operatives and gave fewer other services. During the dispute of the 1960s Neil Blaney proposed to give an expanded DDC a permanent key role in the industry.

The real disagreement was between the farm organisations, following Plunkett, Dr Knapp and the experience of successful exporters abroad, against the Department, the Dairy Disposal Company and the Dairy Products Survey Team (February 1963). Could farmers best manage their own affairs, aided by suitable legislation, or should development continue to depend on direct government intervention?

The NFA provided the essential local support to the IAOS in creamery consolidation (1964) and in organising farmers' meetings in the Dairy Disposal Company areas urging co-operative take-over. Unlike most co-operatives the DDC did not co-operate with the NFA in the collection of membership fees.

Marketing

The disorganised marketing of dairy produce had been criticised by every commentator since the 1880s. The Commission of Inquiry into the Resources and Industries of Ireland (1922) listed faults in quality control and in market development.

The Commission's recommendations for the dairy industry included a farmer controlled National Dairy Council, whose power would gradually be increased. It should administer a National Brand for exports and finance advertising from a levy on production. All legislation should be submitted to the Council before going to the Dáil and administered by the Council when enacted.

In the 1920s the IAOS attempted to carry out these recommendations. The Dutch 'Gerber' system of testing for butter fat was copied, as was the Danish centralised marketing system, with the brand name 'IAOS'. However, Irish law had held that a rule binding members to sell all their product through their co-operative was an agreement in restraint of trade and therefore unenforceable.[12] The Co-operative Agency society could not rely on its member creameries' supply, nor restrict the name 'Irish' to quality tested products. Twenty per cent of creameries opted out of the scheme. Instead of following the Danish or Dutch precedent of legislation enabling farmers to put their house in order, the Department took over all licensing and inspection. The IAOS lost most of its functions. When the Butter Marketing Committee took control of prices and export marketing, creamery management lost all contact with the export market. Managers produced butter for the subsidised home market, though this product had the lowest price in world trade. In fact creameries such as Ballyclough, which had diversified production, were forced to divert their milk into butter to ensure the home market supply. No market was developed for skim. Regulations ossified. It was not possible under the Irish legislation to pay for milk on any basis other than butter fat, nor could creameries switch to more modern testing systems. Payment could not be based on solids non-fat (for cheese), nor bonuses given for cleanliness. According to some creamery inspectors shareholders could not be paid a co-operative bonus above the price paid to non members.

Department officials generally did their job conscientiously and honestly but without market orientation or openness to change.

After 1955, increased milk production presented a marketing problem and opportunity. The Dairy Committee submission to the Greene Marketing Committee was accepted, with the establishment of Bord Bainne replacing the Butter Marketing Committee. Once the necessary enabling legislation to imitate the Danes was in place, Bord Bainne's 'Kerry Gold' development in the British market became a textbook case of good marketing. A premium price was obtained. Had the IAOS been given similar powers when, in 1928, it instituted Associated Creameries with a brand name ('IAOS') and had support of 80 per cent of producers for centralised marketing, the history of Irish dairying would have been very different.[13]

International

In 1967 Irish farmers had the lowest milk prices in Europe. The West German price of 18 pence a gallon, compared to the Irish price of 10.5 pence opened no markets. Growing surpluses everywhere limited access even to the British market.

The NFA had to fill the need for international representation. It was the only body able, or energetic enough, to approach the British Board of Trade objecting to the limitation of imports of Irish butter (1960).

11 International meetings were not too exciting: (l. to r.) Richard Deasy, Michael-Owen Fogarty and Kit Ahern at a meeting in Rome 1957.

When Jerome Buttimer and Sean Healy attended the European Dairy Conference at Cremona, Italy in 1961, Ireland had been omitted from all statistics of the dairy trade; it was not recognised as a dairy exporter. On recent figures there was some justification for this, but they had it changed. The New Zealand Prime Minister accused Ireland of dumping on the British market. No quota should be given to 'non-traditional' suppliers. The New Zealanders could hardly blame the Irish supply of 7,000 tons for glutting the British market to which they themselves supplied 147,000 tons. The position was rectified because of this meeting. It was important that Ireland was in a position to state its case. To be unseen is to be ignored.

The Danes also accused Ireland of dumping. The NFA pointed out that Danish farmers received a subsidy of £4 per cow (total £6 million) and a farm grant of £40 per farm (£11 million); they sold butter at £368 at home, but £305 in the UK (£8m subsidy). If the British market was glutted Ireland could hardly be responsible, supplying five per cent of the market compared to Denmark's 20 per cent. Ireland had the cheapest milk price, so was less guilty of dumping.

To counter-attack, the Dairy Committee lobbied in Britain, arguing that Ireland was a more important market to Britain than Denmark. Denmark bought less from Britain than it sold. The Committee approached the Irish farm machinery importers and joined with them to visit the British Farm Machinery Manufacturer's Association, pointing out that Ireland could only import British machinery if Britain bought from Ireland. The British manufacturers undertook to lobby the British cabinet on the Irish behalf.

The Irish quota was increased from 4,000 tons for the first six months of 1961 and to 12,000 tons a year. By 1969 the quota was 26,000 tons. In these negotiations the Department of Agriculture and the farm associations made common cause to save the last toe-hold on the British market, so carelessly lost by Irish political and technical mismanagement.

PIGS AND BACON COMMITTEE

Pork was exported from Ireland as far back as 1623. The bacon industry, developed for preservation of meat, was especially suited to export and for ships victualling in Cork, the last stop before America. In this technology Ireland was advanced. The Irish firm, Denny's, set up the first Danish bacon factory.

The trade failed to develop its full potential. Protection of grain production added to Economic War tariffs on export almost killed the export trade.

During World War 2 quality standards were let slip. Fat bacon, often the result of feeding with potatoes, found a ready home market, but one which vanished when the consumer had a choice. Irish exports of pigs, live and dead, totalled £300 in 1945. When Ireland awoke, Danish exports to Britain were ten times the

Republic's. The six counties of Northern Ireland had more pigs than the Republic. Even so, pig production accounted for 12 per cent of farm output; consumed the crop of 200,000 acres of fodder; employed 2,500 workers in feed compounding and 4,200 in processing.

The industry was fragmented. There were 40 factories, four of them co-operative. The Pigs and Bacon Act (1935) set up a Pigs and Bacon Commission with some powers of regulation but no development function in farm, factory or market. It was not till 1961 that the Commission was given wider powers and the task of improving quality and promoting markets.

When the NFA Pigs and Bacon Committee met on 4 July1955, its objective was stability with increased production. The fodder question was handled jointly with the Grain Committee from early in 1956.

Energetic marketing to absorb the increased production expected was special concern for this committee. A scheme was presented to the Minister, James Dillon, who saw no reason to discuss the matter with NFA. The Department again fixed the price of Grade A pigs without reference to producers.

The Pigs Committee met the Bacon Curers' Association, as a letter from the Minister had suggested. After two meetings agreement was reached on a joint policy. It required a statutory board:

- Representative of farmers, curers and the Minister.
- Purchasing all pigs for allocation between factories.
- Fixing prices of grade A and B bacon.
- Employing graders to apply common standards in all factories.
- Collecting a levy on all pigs – the proceeds to be used to promote exports. An export quota should be set for factories.
- Working with the Department and the Agricultural Research Institute to improve breeding by progeny testing and other means.
- Acting as adviser to the Minister on trade agreements.[14]

A joint deputation of NFA and curers presented these proposals to the Department in 1958. From then on the right of the organisation to speak on the pig-producers' interest was recognised.

Breeding

The Committee reported that no breeding stock was available in Ireland from which a guaranteed high proportion of grade A bacon could be produced. It was asserted that only 10 to 12 per cent of Irish pigs were making Grade A Special.[15]

Under the Livestock Breeding Acts the Large White breed was favoured to the exclusion of all others. The object was uniformity. The problems of the dual-purpose Shorthorn cow were repeated. Licensing of boars was by live inspection. There was no progeny testing or assessment of carcass. The Irish Large White herd-book gave 17 visual points for judgement, for example, shape of head, set of ears and tail, colour and texture of skin and hair. These were in no way related to market demand or farm efficiency. The RDS had failed to improve quality because prizes were awarded on visual inspection, based on the herd-book criteria.

Other countries had solved their breed problems. When Denmark's export to Germany was shut off in 1895 their type of pig was unsuited to the British market. They set out to invent a suitable breed of bacon pigs by progeny testing. The Committee studied the achievements of the Danish (co-operatively owned) progeny testing stations. The Landrace pig had been tailored and standardised to meet the requirements of commercial production – light in the shoulder, placing most weight in the more valuable ham and extra length in the side.

Sweden began to copy the Danish methods in 1926. By 1930 they had 30 progeny testing stations. The Netherlands followed suit in 1930.[16]

Ireland continued to rely on visual inspection. The first break-through came when 'carcass competitions' were introduced by Macra and the bacon curers. Stock were first judged visually, then after slaughter. Prizes were awarded for accuracy in judgement of the market value. The Department began a pilot scheme of progeny testing at Ballyhaise in 1954, copying the Danish stations of 1904. Farmers and bacon curers were repeatedly refused any say in the running of the Irish stations. The best offer was for a representative of the Committee to inspect the pilot station at Ballyhaise. The Department soothed the Committee with promises of three more stations, but only that in Cork was operating four years later. At the same time tens of millions could be found for the Nitrogen factory. The Committee constantly complained that 'time marches on and we are being left behind'.[17]

Whatever the *potential* of the Large White breed, there seemed no prospect that it could be brought to comparable standard with the Landrace in reasonable time. Importation was recommended as the best way of making up for lost time. Refusing to use Landrace in favour of Large Whites was like refusing pneumatic tyres because lorries had solid tyres in 1900. The Department began limited imports from the UK in 1958.

There were two impediments to importation. The Danes prohibited the export of breeding stock; the Irish Department would have any unlicensed boar's throat cut. Consequently importation occurred in a haphazard manner. Some Swedish Landrace were imported to Britain in 1954. A few filtered through Britain to Ireland but this unofficial, haphazard importation gave no guarantee that the best strains were bought. Farmers paid a premium for Landrace pigs but there was no adequate selection. If the ears cocked forward, farmers would pay the premium. To improve the standard the Landrace Pig Society, with its herd book, was set up in NFA House.

Gerry MacElligott was prosecuted for possession of Landrace. The Court sharply criticised the Department for bringing the case but, forced to find him guilty, fined him five pence. Harry Meredith probably imported the first Landrace boar. He placed it in an open trailer at the door of the Department in Dublin. The staff all passed out to lunch but not one noticed the difference.

There was confusion in expert advice. Dr John Hammond, invited to report on the industry by the Minister in 1955, judged the Large Whites seen in farm and factory 'better and more uniform than those in Britain'. This was taken as an endorsement of the traditional Department policy, provided further improvements were made through progeny testing.[18] Unfortunately, while the British could survive with such standards in their protected markets, the Irish faced world competition which would not tolerate second best.

Fear of disease was a further argument against import. Rhinitis, a relatively minor infection distorting the nose, was found in Landrace. The Irish authorities claimed the disease was unknown in Ireland. Mr Dillon, as Minister, said that he had been 'unable to get any single vet in Ireland, Northern Ireland or England to recommend the import of Landrace pigs. It appeared that the Rhinitis might take a mild form in British but an acute form in Landrace, which acute form could be transmitted back to the native pigs'.[19]

Dr Greene offered to present to the Department one Landrace boar from the Channel Islands, where no rhinitis is found and specialised farms reared stock under aseptic conditions, guaranteed free from rhinitis. Some scientific investigation could then have commenced.

Dr Greene's offer was refused. No tests appear to have been made till the *Farmers' Journal* sent Irish Large White heads for inspection to a laboratory in the Isle of Man. These Irish pigs had rhinitis, undiscovered by visual inspection because the heavier head of the Large White prevented the nasal twist which was the symptom of infection.

The Danish farmers had undertaken progeny testing on their own initiative. Why did the Irish not do as much?

For the Danes the expense of a breeding centre could be justified by a clearly seen profit. A factory, usually co-operative, saw the pattern of demand on the market and knew the type of beast which would satisfy the customer. Having found or bred the right stock, the factory could issue them to its suppliers secure in the contract that all progeny would return when ready for slaughter. Development expense would be met from the higher market price. In Ireland, for pigs as for butter, a contract guaranteeing a co-operative exclusive rights over members' sales was unenforceable at law as an 'agreement in restraint of trade'. In Ireland the improved pigs would be sold on the open market and the progressive factory would gain no advantage from its initiative.

Co-operative pig development units were formed to improve breeding and gain economies in marketing. They were also a response to the fear, prevalent at that time, that large firms would dominate the industry through 'vertical integration' of suppliers tied by contracts.

Marketing

Breeding accounted for only a part of the price premium of Danish over Irish bacon. Denny's may have shown the Danes how to make bacon but they could now show the Irish how to sell it.

Thirty-eight factories, each with its special, secret cure, competed to sell 'Irish' bacon. Even if all were superior to Danish Lur Brand the effect could only be disaster. No grocer or housewife could be sure what the next 'Irish' purchase would taste like. The Danish factories had established a standard product, agreed by co-operative and private producers alike.

Irish bacon, not being a uniform commodity, could not be advertised effectively. Furthermore, the cost of an advertising campaign was impossibly high when related to the few tons exported.

Perhaps the key failure was the lack of continuity in supply. Irish export was a casual outlet for any surplus from the home market. Since pig production is proverbial for its fluctuations, the residual export market volume varied even more than production. When pig deliveries fell from 1.3 million in 1954 to 0.9 million (31 per cent) in 1956, exports to the UK fell from 251,000 cwt to 94,000 cwt (63 per cent). It was impossible to build a network of dedicated agents when they could not depend on uniform taste, regular supply or advertising back-up.

Irish bacon was of secondary interest to marketing agents. A Danish agent in Britain handled 200 tons a week; Irish agents averaged 15 tons. The Danes had 24 agents to handle 5,000 tons a week, the Irish at least 38 to handle 500 tons. Some of the Irish agents owned factories in other countries, to which they owed their first loyalty. 'This leads to lack of confidence in our methods and results in weak selling and the acceptance of unsatisfactory prices.'[21]

In the internal market things were no better. Bonhams reared by the small farmers of Cavan, Monaghan or West Cork were sold to fatteners. The breeders had no contact with the grading of the final market, so their only interest was in large litters and good milking. The producers of fat pigs were also out of contact with market demand. They sold live to dealers at markets rather than direct to factories. In consequence the farmer did not know the grade, could get no bonus for quality. Even a co-operative such as Clover Meats gave preference in price to dealers over members supplying direct to their own factory. Smuggling reached huge proportions, the flow changing direction with each shift in the market. Sometimes 'illegal immigrants' were 10 per cent of supplies. No wonder that Irish bacon exports to Britain in 1956 were only 17 per cent of those in 1938.

The Pigs and Bacon Marketing Board

The Pigs and Bacon (Amendment) Bill of 1960 was to reform marketing on lines suggested by Juan Greene's Marketing Advisory Committee Reports, with a fundamental difference. The NFA had proposed a board with four elected producer representatives, three curers and one from the Department. This would give an effective producer majority when a co-operative bacon factory was represented. However, the new Board had two representatives for about 120,000 producers and three for the 30 curers. This continued traditional government policy that producers should confine their interest within the farm gate. In the 1930s the Bacon Marketing Board had only curers.

The NFA had for four years been recognised as representing farmers at the regular joint meetings with curers and Department. However, Mr Lemass, as Taoiseach, was firm that the curers should be in a majority on the Board. The two producer representatives would not be from NFA or elected but chosen, by some undisclosed process, from Muintir na Tíre, the General Council of Committees of Agriculture, Macra na Feirme, Irish Country Women's Association, Irish Creamery Milk Suppliers and the NFA. The ploy was that used in the drafting of legislation on the Agricultural Institute, to divide and rule. The producers' associations elected Michael Delehanty, chairman of NFA Pigs and Bacon Committee and Dan Moynahan, ICMSA, of Cork, but the combined organisations requested future election be by direct vote of producers, the system favoured by the NFA for all boards.

The North of Ireland Pigs Marketing Board had 14 members, of whom 11 were elected by the producers, the remainder being appointed by the Minister and the chairman was a producer.

The Government ignored the new board when introducing new, stricter grading systems in 1962. The complaint was not against the much needed raising of standards but that inadequate notice was given and so farm income would fall. The grading was revised following a meeting of the president, Rickard Deasy, and Michael Delehanty with the Department.

The Board quickly achieved improvements. By 1963 the proportion of pigs grading 'A Special' had increased from 11 per cent in 1960 to 20 per cent and a further 41 per cent reached Grade A. Agents in Britain were reduced from 38 to 18. Advertising was undertaken jointly by the factories. The export of pork to Norway and elsewhere was opened. Lectures were given all over the country on quality and a technical manager, with two assistants, was appointed to supervise quality control and give guidance in the factories.

Fluctuations in supply over the four-year 'hog cycle' had been the subject of continual complaint from the market and from the factories, which could not fill their capacity in times of shortage. Small producers would move into pig production when the prices were attractive, always coming to market when the glut depressed prices. (Hence the Monaghan farmer's advice to his son, 'when everyone else runs let you walk; when everyone else walks let you run and you will make money in pigs'.) As large specialised producers took over the trade production stabilised. In 1960 110,700 holdings had pigs, of which only a quarter had 10 pigs or more. The number of producers almost halved between 1965 and 1973, though pig numbers increased 10 per cent.

Development of specialised pig farms weakened the solidarity of farmers within the IFA. Having only one product the Committee tended to lose interest in the discussion of general farm topics at National Council meetings. There were often complaints of their non attendance. As in other countries the specialised committee became almost a separate association. The growing importance of specialised producers' associations in Denmark and the Netherlands was described by Flemming Just and J. Frouws, respectively, at a Society of Rural Sociology meeting in Prague in 1994.

The Cartel

In October 1957 an invitation was received from the NFU to attend a meeting of non-governmental producer organisations supplying the British market. The meeting in London of British, Danish, Dutch, Swedish suppliers was the first of many.

The British explained their concern that the oversupply of bacon was against every supplier's interest. Rather than applying for governmental restrictions to protect the market, they suggested a 'gentleman's agreement' to restrict supply – backed by the threat of quota restrictions if their guests did not behave as gentlemen.

The Bacon Consultative Council was formed to regulate the supply of bacon on the British market and so ensure stable and profitable prices for producers.

Figures were produced by Asher Wingarten, Economist of the NFU, showing that an increase of one per cent in supply during any week caused a fall of several per cent in price, causing a reduction in farm income. The optimum quantity to supply could be calculated for each week in the year, with reduced demand during Lent and increases to seaside areas during holidays. Since bacon took six weeks after slaughter to cure, the quantity coming on offer could be known well in advance if information was pooled. The delegates went home, promising to meet every month to give the required information on expected deliveries. Figures they sent indeed – and 20 per cent more bacon than the figures indicated. Each factory had decided that, if everyone else was cutting back, they could increase supply to avail of the higher price.

It looked like a disaster when they met again. Slowly a group spirit was developed by convivial meetings in different capitals. Friendships were made over generous hospitality at dinner and singsongs after the meeting. Delegates became ashamed to produce figures, so obviously false, before their friends. Eventually an effective voluntary cartel was in operation, meeting every month to allocate the bacon supply that the market could absorb at a reasonable price. Three meetings were held in Dublin.

The limit to international co-operation was reached when a joint advertising and bacon promotion scheme was proposed by the Danes. The NFA could not agree. If the housewife bought more bacon the consumption of beef would fall. Denmark supplied 47 per cent of the bacon, Ireland five per cent. In fresh beef the Irish dominated and the Danes did not supply. Our interests ceased to be the same.

WOOL COMMITTEE

Wool sold directly to Dublin through the Tir Connail Co-operative in Newtown-cunningham made 25 per cent more than with the local merchants in 1952.

Macra, with IAOS and Coras Trachtála, held a conference on wool in Ballinasloe in 1953. Attendance was so poor that the platform speakers outnumbered the audience. With courage in convictions rather than numbers a resolution recommending the formation of a Wool Board was passed. Macra appointed a National Wool Committee.

The Committee found an unfortunate position in the trade. Sheep were commonly branded with tar, or with markers which formed paint when mixed with oil. In either case the dirt spread through the fleece during processing. Farmers sold ungraded. Small farmers, who had to sell early to their 'gombeen' grocer in part payment of their account, received 1.6 pence to 2.5 pence a pound (10 per cent) less than those who could afford to wait six months. Dick Langford,

12 Delivery of wool to auction organised by Collinstown, Co. Westmeath, NFA

of the IAOS, arranged that the Scottish Wool Growers co-operative invite a Macra group[22] to inspect their marketing arrangements. Scottish Wool Growers offered to buy the whole crop on a graded basis.

When the NFA was formed in 1955 the work of the Wool Committee continued and the first sales were held in that year with Messrs Fyfe Davies. In following years Irish Wools, joined later by other firms, operated the scheme. Sales took place at 100 centres in 23 counties during 1960. Farmers guaranteed a minimum of 9,000 lb of wool at each centre. The quality bonuses, mainly based on absence of 'deleterious matter', added 1.3 pence a pound on about one million pounds sold, a gain of about £19,000.

These developments were unwelcome to the established merchants, who at first met the new entrant with unfair competition. On days of wool sales in Kerry shopkeepers set up buying points on the farmers' way into town, offering one penny a pound, without grades, above the going price for small lots. They found no takers. Later the local merchants adopted grading and paid competitive prices. In this way the benefit was spread over the whole output.

The Macra Sheep and Wool Committee was transferred to NFA in 1955. Sheep numbers had doubled in the previous ten years and the collapse of lamb prices in that year presented an acute problem. The market fell in sympathy with that in Britain, though only three per cent of product was exported. Meetings were held with Irish Fresh Meat Exporters and a levy on producers was proposed to form a stabilisation fund which should be aided by the Exchequer.

Serious disease problems – an incidence of 95 per cent liver fluke infection and the prevalence of Scab – deterred British importers of live sheep. This was handled jointly with the cattlemen in the Health and Disease Sub-Committee.

POULTRY COMMITTEE

At the turn of the century ten chickens gave the income of one cow. This belonged to the farmer's wife, jealously guarded from the co-operative lest it go the way of the creamery cheque to the husband's pocket. The trade did not prosper. Government grading schemes (1924) raised quality, gaining a premium in the market, but in the 1930s tillage protection made fodder dear and Economic War tariffs made export impossible. Dr Bob O'Connor calculated that 'every hen died in debt'. Poultry numbers fell from 23 million in 1930 to 20 million at the outbreak of World War 2, and 17 million by its end. In contrast, imported feed and guaranteed prices induced such growth in the North that the poultry flock became 10 per cent larger than in the Republic.

After the Trade Agreement of 1948, James Dillon promised to 'drown Britain in eggs'. Poultry numbers rose to the 1930 level and exports reached £7.5 million in 1952. The 'Deficiency Payments' system by which the British farmer was guaranteed an economic return regardless of the market price killed the export trade. British production rose to glut their home market. At one stage the British farmer received 17.5 pence a dozen for eggs, which were sold as surplus to Germany at 12.5 pence, Irish exports fell back to £600,000 a year.

Confined to the home market the seasonal price fluctuations became extreme, varying from 7.5 pence a dozen in the Summer to a Winter price of 30 pence. The traditional small farmers' wives, selling most of their output in Summer, were driven out of the market by the factory producers.

If the Poultry Committee faced a depressing task when it first met in June 1956 things got rapidly worse. Total poultry fell from 16 million in 1955 to 8 million in 1959. Exports fell to £137,000. The Committee concentrated on the home market, seeking to establish a minimum price for eggs and to reduce the seasonal glut.

The Marketing Advisory Committee endorsed the Poultry Committee recommendation of a marketing board exporting only Grade A eggs. However, there were virtually no exports. Consultations were held with government, hotelliers, hatcheries. Visits were paid to the Poultry Section of the Fatstock Marketing Corporation in London, to Smithfield auctions, to conferences in Belfast. The International Federation of Agricultural Producers (IFAP) held a world poultry seminar in NFA House in 1959.

For all this effort there was little to show. The export market was not free. The home market remained fragmented. The seasonal glut disappeared, but so did the small producer.

Two areas of development remained:

1) *Hatcheries* could develop an export trade. Eggs could not be exported but day-old chicks could be sent to produce subsidised eggs inside the protected British market.

2) *Turkeys* Department regulations permitted only the traditional 'Bronze' breed of turkey. They had darker flesh and thinner breast than the improved white 'Beltsville'. Mrs Gallagher, manufacturer of Urney confectionery, smuggled eggs under maternity wear to provide a foundation stock of Beltsvilles in 1952. Macra displayed them at a promotion (paid for by Irish Shell oil company) in the unlikely venue of the fashionable Hibernian Hotel, Dublin. As soon as the Department permitted the import of these improved turkeys production expanded for the home and export markets. The Committee organised turkey auctions, similar to those for wool, which gave fairer prices. A trial was made with 30 auctions in 1965 and about 50 venues a year after that. Carcass-judging competitions, similar to those for cattle and pigs, were introduced at agricultural shows (1958).

The Committee gave evidence to the Tribunal on Cross-Channel Shipping on practical export difficulties. Jim Quinn, secretary of the Committee, followed the turkeys to Smithfield market with his camera. Turkeys were manhandled onto the ships in loose crates. In the course of this laboursome task some crates always dropped and spilled their contents – a source of Christmas cheer to the dockers. Rough handling and temperature changes so damaged and bruised the contents that they were known as 'black and white' turkeys in London's

13 Delivery of Irish turkeys in broken wooden boxes

Smithfield Market. Offered with the salvage produce at the end of the sales, they made the lowest prices.

Refrigerated containers would improve the condition of the stock and prevent pilferage. An NFA group met the dockers' union only to be told the facts of life by the dockers. It took 120 men to load by the old system; 16 with containers. The Christmas turkey trade gave temporary employment to the dockers' children, their spending money for the holiday. Additional employment in farms and factories by a prosperous export trade was no benefit to dockers or their union.

Refrigerated containers were obtained to by-pass Dublin through the port of Greenore.

GRAIN COMMITTEE

Ireland had been Britain's breadbasket during the Napoleonic Wars, protected by a system like that of the modern CAP. A target price was set and import allowed only if that price was exceeded. British industrialists used the Irish Famine as a pretext to abolish these Corn Laws in 1848. In fact the Irish gained little. Most of them ate potatoes, not bread. Cheap imported grain only lowered the value of their cash crop. Irish grain acreage fell from 429,000 acres in 1851 to 21,000 in 1931 as grain from the newly opened plains of North America, Argentina and Australia came on the market.

In the same period Denmark and the Netherlands adapted to the changed market conditions by allowing cheap imports to feed intensive livestock production for export, so balancing their trade at a high level. Ireland chose livestock production based on seasonal grazing, with little labour, low input costs and low output.

During the Economic War (1932–1938) tillage was encouraged in the drive for self sufficiency and to give employment. Grain growing was mistakenly thought of as labour intensive – as it had been before mechanisation. During World War 2 farmers were obliged under the Compulsory Tillage Orders to till 37.5 per cent of their arable land, with up to 10 per cent in wheat. This was applied in a Draconian fashion to areas unsuited to the crop. Even the Dublin liquid milk producers, quite unequipped for tillage, were forced to plough their grazing land. Then scarce fuel had to be expended on transport of milk from unregistered producers in Munster. The wheat acreage was maintained at about 280,000 acres. Fertiliser and lime were almost unobtainable, so application fell sharply and soil fertility was depleted for the future.

In 1955 all European countries except Denmark gave protection to grain growers, continuing wartime efforts for self-sufficiency. World production had regained pre-war level. Canada and the USA had stock piles of grain, which they exported at dumped price. The Irish price for wheat, though the third lowest price in Europe, had been reduced by 15 per cent in 1955. Home production was falling fast towards the 1920 levels.

Wheat was a cash crop requiring less capital than cattle. Since it was not for export the cost of protection would not lie on the Exchequer. The national requirement of milling wheat was about 450,000 tons a year, of which the Irish harvest supplied some 300,000 tons. Storage capacity was 350,000 tons. The NFA Committee argued for a price under the Wheat Order calculated to give this target quantity of millable wheat. Full use of storage would give lowest cost per ton.

The Minister for Agriculture, Paddy Smith, was not hostile to restoring the price to the 1954 level, provided that no export surplus would burden the Exchequer. He arranged an interview with the Taoiseach, Eamon de Valera, to state the case. De Valera was probably the most courteous minister to meet an NFA delegation. He showed a surprising knowledge of the wiles of the market – even the value of weed seeds taken from the wheat and used in animal compounds. At the same time he had instant control of the meeting and Paddy Smith's subservient 'Yes, Chief' showed where the real power lay.

The NFA Grain Committee sent wheat samples to the Swedish research centre at Svalov. Their tests showed millable quality, encouraging the Committee to press for the maximum use of Irish wheat in the grist.

A letter from Svaloff to the NFA read:

> Dear Sir,
>
> I am now able to forward the result of the baking test. Your samples... were bulked and tested. For comparison I obtained a Swedish sample from Wiebullsholm in south west Sweden.
>
> The general outline of the comparison between your sample and ours is that yours is the best one ...
>
> In comparison with our inventory quality of Swedish wheat your sample is definitely better than the average Swedish Quality.
>
> If you can produce a wheat with a bread volume and a protein content like that you sent us, the Irish millers should not have any trouble to produce a acceptable flour of home produced wheat only.[23]

Steps to improve quality began in 1956. A new world of technicalities opened for the growers. The Committee had ready response to requests for samples, grown with varying nitrogen fertilisers, for testing in the Institute for Research and Standards.

A quantity of flour made from 100 per cent Irish grist was milled for the committee by Mosse, millers in Kilkenny. Downes, of Dublin, baked 300 loaves a day for Mrs Lawlor, of Naas, caterer to the RDS Spring Show in 1958. The Army also used the all Irish Loaf without complaint. The foreman baker said that no extra care had to be taken. If the consumer would accept French-style bread a higher Irish content could be used, but the baker's equipment was designed for imported wheats. The conservative consumer resisted, demanding the 'batch' loaf, with soft sides. There was method in the consumers' demand. The 'pan' loaf can contain more water and, therefore, less food.

For confectionery and home baked soda bread there was little need for 'strong' imported wheat. Bewley's Cafes confirmed this by trial.

Millers were not averse to the use of native grain to give profitable use to their storage. Most mills were situated at the ports and equipped for hard (high protean) imported wheat. They preferred to supply the consumer with the product suited to the equipment they had. The farmers wanted to supply as much of the home market as possible with acceptable price and quality. One of the millers' objections to Irish wheat was that it gave a lower volume and weight of bread. In effect more water and air could be sold with the Canadian loaf.

Stable prices had embarrassing success in 1957. The millers bought 380,000 tons of high class milling wheat, a record, but there remained a surplus to be exported at £17.50 a ton. The prospect was that, with improved husbandry, yields would rise by half a barrel an acre each year for the future, so an annually increasing surplus was in prospect. In fact average yields rose beyond expectations, as shown in Table 5:

Table 5 Average Yields of Wheat in Ireland

Years	*Tonnes per Hectare*
1910–19	2.5
1920–29	2.4
1930–39	2.5
1940–49	2.1
1950–59	2.7
1960–69	3.5
1970–79	4.1
1988–92	8.0

Source: *Statistical Abstracts*

The Married Price

The price of wheat was a political decision on which farmers were rarely consulted. Farmers wanted a fixed price but the Exchequer could not bear an open-ended guarantee for all the wheat produced. They might have to subsidise an export surplus. The NFA solution was a flexible 'married price'. A target volume should be set each year for the national crop of millable wheat. Surveys of farmers' intended acreage, of planting and growth should be made during the year to guide the farmer and to establish the probable crop. If a surplus was expected a levy should be collected from all millable wheat delivered at the mills to finance disposal of any surplus for feed or export. In this way all farmers would get the same price; the millers would gain by full use of equipment and the Exchequer cost would be limited. Any surplus in the fund would be returned to farmers at the end of the season.

The Irish Flour Millers' Association and the General Committee of County Committees of Agriculture proposed a quota system to limit production. The ICMSA , in line with their milk price policy, favoured a 'two-tier' system with a higher price for small quantities.

The alternative was a cut in the price for all wheat.

The Government agreed to the married price principle in 1958, but only disclosed later that the 1957 price was limited to 300,000 tons.

The NFA carried out a preliminary survey for growers' guidance. It showed the probability of a levy of 65 pence to 70 pence a barrel. Growers heeded advice to cut back their planting. The levy was reduced to 29 pence a barrel.

The price in the Republic attracted Northern smugglers in 1969, a most unwelcome development when producers paid a levy to dispose of a 70,000 tons surplus of native wheat. The NFA organised farmer patrols to prevent illegal imports over the Louth and Monaghan borders.

An Bord Grain

A Grain Board was set up for a trial period in July 1958 to administer the wheat market. The members were appointed by the Government, but on the nomination of the various associations instead of by arbitrary political decision. The NFA nominated the chairman, John Orpen, and five of the nine feeder and grower representatives; the cooperatives nominated the manager of a co-operative with a large feedstuff trade; traders had three representatives. Sean Fitzmaurice, press officer of NFA, became manager.

The harvest of 1958 was disastrous, quite reversing the expectations of a surplus of millable wheat. All the interested parties, NFA, Millers Association, Department and Grain Board worked together. On 11 September a new procedure was started by which wheat, sound but unsuited for flour milling, was bought at milling price instead of the sacrifice price of £2.50 a barrel or less paid till then. The Board supplied the miller with imported wheat in exchange and sold the unmillable wheat for animal feed.

Disposal of the levy funds, £700,000, was controversial. Farmers wanted it returned, since there was no export surplus. The Board's profit on imports paid for the exchange scheme. Eventually about £420,000 was returned to growers and the price of bread was reduced by a farthing (0.1 pence) a standard loaf.

The farmers' loss had been reduced to £5.5 million. Hardly a victory, but proof of the Board's effectiveness.

The weather was inclement again in 1960. On 8 September the Department found it necessary to write to the Millers' Association: 'I am directed by the Minister for Agriculture to state that he has received reports which appear to indicate that an undue proportion of the wheat of the 1960 crop which has been presented to the mills so far, is being graded as unmillable. This indication is supported by reports of differences in the proportion of wheat being graded unmillable by individual millers even when operating in the same area. It will be necessary for the Minister to consider taking such measures as may be required to give effect to his views and policy as indicated.' On 12 September the market was relieved by a Department Order that all sweet and sound wheat be purchased at Wheat Order prices.

The Grain Committee complained to RTÉ in 1961 that NFA press releases on the harvest arrangements were suppressed. They were mentioned on the 10.15

Evening News on 17 February, but not afterwards. There was no comment on the price reductions or the market arrangements, though the farmers were anxiously awaiting the information to decide on planting.

The Committee used various means of pressure to influence millers' policy. When it came to the notice of the Committee that Rank's mill in Limerick was importing during harvest time and refusing to take Irish grain, the manager reluctantly received a deputation. He told them to mind their own business. Threat of a boycott of Blue Cross animal feeds opened the mill that very day.

On the other hand, Michael-Owen Fogarty, who had a mill in Aughrim, was persuaded to take in a widow's rather sprout-damaged wheat at full price. In a shop afterwards he heard her refusing to buy bread made from his flour: 'I wouldn't touch that stuff! If you only saw the dirt he buys!'

Relations with the Flour Millers' Association were respectful, even when differences of opinion and of interest were lively. Each party knew that their existence depended on the other. The consumer had to get better bread. If there was no native wheat the millers' large investments in storage and drying equipment would be valueless.

On Sean Lemass' advice, the millers employed a public relations officer to establish friendly relations with farmers. He ensured happy meetings by careful selection of invitees from county committees for agriculture and others who would not be too abrasive. No NFA Grain committeemen or other abrasive types were among the guests. We are told that the dinner parties were indeed harmonious.

Quality Standards

Chaucer's miller had 'a golden thumb to judge an oat'. The grower was still dependent on quality standards and market regulations which he half understood and could not check. The Committee was fortunate in its chairman, John Orpen of Wexford. A graduate in engineering from TCD, he gained unusual promotion to commandant in the Artillery of the Irish Army during World War 2. His control of figures and logistics was formidable. He was backed by the committee secretary, Míceál Ross, later prominent in research with the ESRI. With co-operation from the Flour Millers' Association, their rivals in argument, the Committee mastered the complexities of the trade, visiting research stations in Svalov, Sweden, and elsewhere. The NFA was the first Irish farm organisation to hire a consultant chemist, John Corr.

High moisture content is a problem with Irish wheat. In wet harvests moisture could reach even 35 per cent. In other countries grain is dried on the farm. It can then be held in store, avoiding the gluts at the mills experienced during harvest in Ireland. Grain for animal feed could be dried with the same farm equipment. Irish Millers argued that only they had the expertise necessary for drying. The whole crop was transported to the mill, sold at £2 a barrel and bought back at

£2.90. After repeated requests millers agreed to buy farm dried grain at 14.5 per cent moisture in 1960. The NFA welcomed the development but could not agree that the allowance for farm drying and storage should be less than that charged by the millers or their agents.

In 1961 the NFA wanted 3.3 pence per barrel a month for storage on the farm, the same as the millers' agents. A payment of 0.8 pence was conceded. Eventually the matter was solved by entry to the Common Market and adoption of international standards satisfactory to the farmers.

Government price regulations made a crude adjustment for moisture. Under the 1955 Wheat Order price was cut if moisture exceeded 23 per cent but no bonus was paid for lower moisture. It paid the farmer best to satisfy the minimum standard. In fact it would pay a grower to water dry grain to maximise their income.

The millers set standards for grain and did the testing at the mill. In 1960, in co-operation with the Department and the millers, a service was provided by which growers could have their grain retested for moisture on payment of a fee to the Central Wheat Laboratory.

Further quality tests were introduced. The NFA did not object to the raising of standards but actively protested that the regulations should not mean a price-cut. It was unfair that one party to a bargain should fix standards and make the tests, particularly as complaints were made that '90 per cent of grain was accepted as millable in Kerry but only one per cent in Meath' (1962).

The 1962 millers' arrangement on standards and tests was terminated in 1964. The NFA had never liked its standards or methods. Millers wanted the 'Maltose Test'. The farmers objected on technical grounds, preferring the Hagberg test. Dr Olered, a Swedish chemist, recommended the Hagberg Test and a neutral laboratory for appeals and for information; with a consultative committee to discuss and make recommendations. The Flour Millers' Association accepted Dr Olered's recommendations in full. In subsequent years grading became ever more stringent. The farmers' complaint was not against penalties for inferior wheat but at the absence of bonus for quality. Producers rapidly got the message that medium quality paid best and further improvement in bushel-weight or moisture content did not pay. Three harvests with good weather conditions, 1966 to 1969, made progress easier.

Bulk Handling

Grain was handled in sacks in those days. The millers allowed their agents 5 pence a sack on all wheat. If a farmer provided his own sacks, or delivered in bulk, the saving accrued to the agent. Few farmers were even aware of the situation but were happy to accept the small benefit when the millers conceded them similar payments.

The Grain Board was dissolved in 1973 in preparation for entry to the Common Market. Government monopolies could not continue. It was regretted that this organisation, which had stabilised the market and improved standards by co-operation between millers and farmers, was not continued in co-operative form like Bord Bainne or the Potato Marketing Company.

Malting Barley

The Beet Growers' Association had satisfactorily negotiated contracts for sale of malting barley to Guinness, at a premium price, for many years. It was desirable to co-ordinate all sections of the grain market. Farmers not producing beet required representation. From 1956, at the request of the Carlow beet growers, negotiations were conducted by a joint NFA/BGA team.

Fodder Barley

James Dillon introduced high yielding fodder barleys after the War. A great increase in acreage resulted. Past experience led men to expect a fall in price at harvest. The Grain Committees of Cork, Waterford, Tipperary and Kilkenny, the main growing areas, tackled the problem by co-operation between Southern co-operatives and those in the feed-buying areas of Cavan, Monaghan, Leitrim. By guaranteeing a price of £2.15 a barrel at 20 per cent moisture the standard for the whole market was set. The Minister helped by limiting imports during the harvest weeks.

Pig producers must pay for the barley. The background to the agreed price for barley was the joint report on the Marketing of Pigs and Feeding Barley (March 1956) from the two committees. In Canada the Government did not fix fodder prices. The pig producers' association met with the grain growers under the auspices of the Farmers' Federation, agreeing a compromise prices binding on all. In the absence of such agreement neither cooperatives nor Minister could have adopted a scheme acceptable to both parties.

Acreage grew from 75,000 in 1954 to 110,000 in 1955 and 208,000 in 1957 because improved varieties were introduced. It was necessary that there should be a co-ordinated policy covering all grains, but the Grain Board's function was limited to wheat. Irish grain prices were low by international standards at £18.5 per 100kg. compared with £19.66 in Britain; £23.95 in France; £24.5 in Netherlands; £32.1 in Germany.

In 1958 imports continued while the Board still had supplies of native wheat for feed. In December the Department introduced a regulation restricting licences for imports of coarse grain and offals to purchases of fodder wheat in the ratio of 2 to 1. Maize imports were used to control the price of fodder barley and unmillable wheat. Imports were made through the Board, which released the grain to those compounders who paid the farm price of £1.90 a barrel. In this way the delivered price throughout the country was set. The poultry and pig feeders in the small-farm areas of the North and West were placed on an even footing with the grain growing areas.

The NFA supported the IAOS efforts to set up co-operative compound feed mills, gaining a dominant position in the market.

Oats

Oats, the product of acid soil, was particularly grown on subsistence farms and in the seed potato area of Donegal. A small part of the crop was sold for horses and as breakfast porridge. The liming of soil and the decline in the demand for horse fodder, reduced the acreage. The advent of Kellogg's Corn Flakes reduced the market for porridge meal to a negligible quantity in a decade. Such is the power of intensive advertising.

Some success was gained in developing a market for oats in the West Indies for race horses, and in regulating imports of oats for bloodstock in Ireland. In practice nothing could be done in a dying trade.

VEGETABLE COMMITTEE

The Potato and Vegetable Committee met frequently, under the chairmanship of Tom Brangan, of Dublin.

The first success was the Voran potato case. Cemici Teo, under their Dutch manager, imported potatoes of a Dutch variety, Voran, because of its high protein content. When propagated in the seed producing areas of Donegal, Aphid infestation appeared in neighbouring crops, causing loss to farmers whose certified crops were condemned for use as seed. Individual claims for compensation were refused. The NFA interviewed the manager of Ceimici Teo, whose attitude was simple. He did not deny responsibility. However, he said, the loss of each individual was too small to justify the expense of a law case and its appeal to a higher court. Therefore, he refused compensation.

The NFA response copied that of the Swedish Farmers' Union (RLF) in dispute with their electricity suppliers. The NFA collected evidence at local meetings in all affected areas, selecting particularly clear-cut cases from each. These, and other, growers subscribed to legal expenses, authorising NFA to handle their complaint. A court order for disclosure of documents was obtained against the company.

The next meeting the company saw the change in the balance of power and paid up.

Seed Potato Exports

The Department had done excellent work on the development of stock and in quality control before Independence. Ireland was a leading exporter. Subsequent developments have been much criticised on technical grounds. Exports fell as other countries overtook Irish seed. Table 6:

Table 6 Production and Export of Irish Seed Potatoes

Year	*Area Grown*	*Export*
1939		30,000 tons
1950	14,700 acres	30,464
1955	13,075	26,662
1966	12,664	37,185
1973	9,006	24,165
1992	2,852	6,669

Source: Irish Potato Marketing Ltd

To restore exports after the war, the Potato Marketing Company was established in 1950 by potato merchants. It had a legal monopoly of all potato exports, except earlies, till 1973, but no function on the home market. Its record was bad. Acreage in potato production continued its rapid decline from 800,000 acres a year to 170,000 in 1965 and stablilising at 40,000. Exports peaked in 1969 with 53,000 tons. In spite of natural advantages, potato exports, except earlies, stagnated while those of Denmark, Netherlands, Germany grew rapidly.

A prime cause of loss of markets was failure to develop new strains. The company was aware of the need but the legislation necessary to protect breeders' rights was not introduced until 1980. In the wasted years many varieties which could have been obtained on franchise for Ireland were lost. The traditional varieties, such as Arran Banner or King Edward could no longer compete. From 1969 IPM entered into agreements with the Agricultural Institute, from which came three new varieties, Cara, Red Cara and Avondale. The Dutch list has approximately 300 varieties.

To make matters worse, there was a severe deterioration in tuber health standards, resulting in financial claims and loss of confidence in the market.

Growers were dissatisfied with the farmers' representatives appointed by the Minister. They tended to appear later in business as merchants. All licensed growers were registered, so election of growers' representatives, the practice in more successful countries or in the BGA, would have been easy. In the absence of growers representatives there was a failure to communicate information on markets and the varieties demanded for export. Transport was described as 'haphazard'. Research and development of new varieties was inadequate. The Company had no responsibility for the home market, where supplies were sometimes inadequate, and there was no system for disposal of occasional surplus. Shopkeepers abused possession of their seed quota, refusing to give seed sacks unless farmers bought other supplies, regardless of price, from their store. The English farmers' co-operatives expressed their satisfaction with Irish seed quality, and offered to buy their supplies from Ireland, provided they came from known co-operatives. The offer was rejected and the trade lost. The Company, using its position as monopolist, refused to allow this international co-operation of farmers. Louis Smith attended and reported on the meeting in Liverpool, representing IAOS.

There was no structure for developing export of early potatoes.

The Potato Committee Report of 1957 records that their meeting with the Company was fully negative. In particular a letter was received, stating that the election of farmer representatives could not be agreed 'as the PMB would have no control over the farmers' representatives on the Board if they were selected by the growers'. The Board's reply was passed to the Minister. Apparently he agreed their policy on farm representation, since no action was taken.

In 1961 the Minister offered a place on the Board of the Company to Neville Chance. Instead of accepting the Minister's appointment a vote of producers was held. Neville Chance was elected. The Minister refused to make the appointment on these terms. Dr Greene's approach to the Taoiseach and NFA threats to publish a strongly worded statement on the Government's rejection of democracy brought Chance to the Board as NFA nominee. The farmers, who wanted 50 per cent representation, were partly satisfied.

In due course the Board became a co-operative to conform with European Community regulations.

Co-operation with the Northern Irish seed potato exporters was urged by the Committee. Price-cutting competition in marketing a similar product under the name 'Irish' was hurtful to both areas.

The seed potato trade has almost died of neglect in spite of natural advantages.

Marketing

A request for representation on Córas Tractála, recognising the importance of the export of potatoes and the opportunities for expansion in vegetable exports, was refused.

Noting that ships were travelling empty to the West Indies, a trial consignment of vegetables and oats was consigned, with the co-operation of Coras Tractála. The sale was successful but the Dutch shipping company was unwilling to continue.

Active support was given to co-operative groups for washing and packaging ware potatoes to the standard found abroad. Special rates were negotiated for shipping earlies.

The Irish housewife was concerned at the presentation of dirty potatoes sold in packs which prevented inspection and concealed tare. There was no certification of quality or education on varieties. The farmer gained no quality bonus – lost nothing when soapy potatoes were produced with over-application of Nitrogen. A handful of clay in the bag for sale cost the farmer little.

Action was taken, mostly at local level, against the dumping of potatoes from Britain. These were bought from farmers by their Marketing Board for £15 a ton in 1965, with a farmer's option to repurchase at £2.75 a ton, provided they did not reappear on the British home market.

Again, contact with the Common Market belatedly raised standards of grading and presentation.

Fruit, Vegetables and Tomatoes

In contrast with the potato experience, the grading of tomatoes was a model exercise by the Tomato Growers' Association, a subsidiary of the Agricultural Association of Ireland.

Tomato growers were protected by a ban on imports during the Irish season. The length of the prohibition period was negotiated with the Department annually. Large quantities were brought in before the embargo was placed. A curious price pattern emerged. Tomatoes were cheap (but less tasty) when import from the Netherlands was open and again at peak Irish production. The grower made his money in the shoulder periods at each end of the Irish season.

From 1956 the NFA joined with the Irish National Fruit Growers' Association for regular price-fixing discussions with the Jam Manufacturers.

The horticultural advisers of county Wexford, led by Chief Agricultural Officers John Murphy and Michael Connelly, saw little point in encouraging soft fruit production and quality if the result would be a glut on the home market. They acted against Department policy by taking the initiative in seeking export outlets. Supply contracts for the whole output were then signed by farmers. A tour of Wexford trying to tempt farmers to sell black currents outside the contract showed that, regardless of price, none was available.

This example of integrated work in production and marketing by the advisory officers shows what might have been accomplished had the Department shown the positive approach to farming and its integration with marketing found in competing countries.

The Dublin market served the whole country and did it with exceptional inefficiency. Country town markets were so small that producers sent their vegetables to the Dublin market. Shops, often from the same locality, bought their wholesale requirements in the same Dublin market.

Market organisation was medieval. Ungraded produce was brought into the market building on payment of tolls to the Corporation. There many agents and auctioneers conducted a mixed wholesale and retail trade on commission. A householder could bid equally with a multiple store, take what was wanted from the lot, and leave the rest to be auctioned again. Congestion of traffic round the market imposed a heavier cost in wasted time than the visible cost of tolls and fees.

Where better, more stable prices and an auction fee of under one per cent was usual abroad, the cost in Dublin was uncertain but estimated at 20 per cent.

Because there was no grading, the purchaser had to inspect each container, presuming that the best selection was on top to conceal the lower quality below. Most serious was the isolation of the market. Elsewhere in Europe, where produce was graded and quality guaranteed, dealers could stabilise prices by off-loading a local surplus to another area.

The supermarket chains, which came to dominate the market, wanted regular, graded supplies even more than cheapness. They would not waste time in the market. An effort to reorganise the market co-operatively, with graded products and wholesale auction procedures, were made by North Dublin Growers, for which the NFA canvassed capital subscriptions. The firm failed through lack of member loyalty. It was too easy for the individual near the city to sell what he could privately, undercutting the market and then sending the residue to the co-operative.

AD HOC COMMITTEES

Topics of interest to some group or important for a short emergency were handled by temporary committees. A few of these were of interest.

Game Committee

Game rights were transferred to farmers from the landlords in 1958. This left a control vacuum. Neglect would allow vermin to increase. The proposals of the gun clubs would multiply the number of guns and infringe the farmers' potentially valuable rights. NFA supported the views of the St Hubert Club of Ireland at a meeting in Cahir, May 1956, for regional game clubs on the model found in Kildare. The Ministers for Lands and for Justice promised co-operation.

Itinerants

Itinerants became a problem after World War 2, partly due to an influx from Britain. Wartime restrictions and conscription were particularly onerous to nomads. At the early Executive meetings complaints were heard of wives intimidated by beggers when alone on isolated farms and of damage to property by itinerants' livestock.

A Government commission considered the matter in 1961. The NFA formed a committee to prepare a submission. Comments were sought from branches and reports obtained from sister organisations in eight countries.

The Committee found that, unlike some other countries, Ireland's itinerants were not a separate race or culture. Formerly they had followed a useful and respected skilled trade as tinsmiths repairing buckets, kettles or milk churns. Their work as traders and casual harvest labour made them welcome to farmers. There was no prospect of similar employment in the future.

The settled community had a responsibility to protect the rights of itinerants' children to health, education and acceptance into society. This was not possible with an itinerant way of life. In short, while better camping sites might alleviate poverty, they were likely to become a poverty trap, perpetuating the growing social problem.

The NFA submission proposed a redundancy and retraining package of the type given to small farms in Sweden. While adult itinerants might be unable to change their way of life and, therefore, could be helped to a reasonable standard of comfort in it, the children should not be condemned to poverty. They should be assisted to become ordinary citizens. Settled housing and education enabling them to earn a living were the proposed solution for the rising generation.

Diversified Output

Alternative crops attracted attention. The Forestry Committee had a discouraging task. The prospect of a fortune from the timber crop forty years hence has little attraction to the average farmer, fifty years old and pressed for current income. Visits to the Department of Lands[24] met little enthusiasm. The official view at that time was that state forestry was more efficient than private. Forestry was regarded as a way of using land unsuited to agriculture. Encouragement of private forestry would raise the price of such land against the Department.

Agri-tourism was immediately attractive, but not a dominant interest. The obtaining of game rights (retained by the old landlords under the Land Acts) and the development of hunting was encouraged.

Horse breeding was a traditional skill which merited the attention of a continuing committee.

Consideration was given to many crops: maize, rape, bamboo, herbs; and to livestock such as deer, ostriches and goats.

14 Modern milk delivery

6

Central Committees

Commodity committees, directly involving members on the farm, at branch level and through commodity associations, have a high profile. Committees working in head office are usually unseen. None the less, their contribution to the individual farmer's welfare has justified the existence of a unified central association many times over.

AGRICULTURAL INSTITUTE COMMITTEE

This was particularly evident when, unexpectedly, funds for research became available from the United States in 1954. Macra had direct knowledge of the unsatisfactory state of agricultural research and education in Ireland. Research was of such prime economic importance that it was a priority for the founders of NFA. Scarce funds were expended on publishing a report when the Association was as yet in embryo.

On 1 April in 1900 the Department of Agriculture and Technical Instruction took over the work of research, agricultural education and the advisory services of the Royal Dublin Society, together with its animal breeding services. The objective was to unify all the services under one command. From 1926 the Faculties of Agriculture in universities were funded by the Department of Agriculture, not by Education. The advisory services of the County Committees of Agriculture were under the Department. When the Department of Education founded vocational schools in the 1930s agriculture science could not be taught in them because that would trespass on Agriculture.

The monopoly stultified progress. Merrion Street's orthodoxy lacked the stimulation of outside ideas. Liaison within the monopoly structure was slight. Universities, for example, had no direct contact with advisory services or farmers. Neither farmers nor the subordinate institutions had direct influence on Department policy. Consultative Councils, where they existed, met rarely or not at all, as described in chapter 1.

Several government commissions had noted the unsatisfactory state of agricultural education. The Drew Commission (1926) led to the foundation of faculty of Agriculture in UCD and of Dairy Science in UCC, but starved research. The Committee of Inquiry into Post Emergency Agricultural Policy (1945) stressed the need for agricultural teaching in the vocational schools. It was without effect. Outsiders, such as the OEEC experts in their report of 1950 (in its published version), praised the advisory services while pointing to their inadequacy. Ireland had one adviser to 2,000 farms; Denmark and the Netherlands one to about 300. Funding for research was negligible, as was the specialist back-up.

The OEEC Report of 1950 commented on technical training: 'Deficiencies are apparent in Agricultural Economics, including training in marketing, farm management, co-operative organisation and agricultural policy.' Among other areas criticised were Agricultural Engineering and Adult Education Methods. Ireland spent £500,000 on tourism but only £69,530 on agricultural research in 1950. This compared with Denmark's £230,000 official and substantial private agricultural research funding. Denmark had 515 agricultural scientists advising 208.000 farmers; Ireland 150 for 250,000 farmers; their agricultural colleges trained 2,500 young people a year compared with Ireland's 300. In real terms the grant to the Faculty for Agriculture in UCD had declined 30 per cent compared with pre-War, though the number of students had multiplied by four. Laboratory space had in some cases been reduced below that available in 1911. No wonder Danish produced three times the Irish output per acre. The experts co-opted to NFA commodity committees had been scathing in their comments on Irish agricultural research.

The Research Institute

The United States took the magnanimous and foresighted step of helping the devastated combatant countries of both sides in World War 2 to rebuild their economies. As a neutral Ireland was debarred from Marshall Aid, but was able to borrow dollars to buy the scarce goods only available in USA.

A condition of these purchases from the USA was that the goods be carried in American ships, whose charges, paid in scarce dollars, were uncompetative. This caused no concern to the countries receiving gifts, since the American taxpayer met the bill. Ireland, which had to pay for the shipping, was refused a refund of the £6 million excess cost. The Americans did something better. They made the money available in a special Grant Counterpart Fund for agreed development expenditure. It was a most fortunate decision. Instead of importing maize as a short term benefit to pig producers, £1,840,000 was made available for a new Research Institute by the agreement of 17 June 1954.

The Department of Agriculture White Paper proposed further centralisation. Higher education should be taken from the universities and concentrated in a new agricultural university. The Minister should have full control with powers 'to alter or expand research programmes', to appoint and remove the Director.

The NFA and Macra saw little advantage in extending a system which had given such poor results in the past.

In January 1955 Macra set up a committee to advise its Council on the proposals for an Agricultural Institute. This was expanded to a Joint Commission of Macra and the NFA. Patrick Quinlan, Professor of Mathematical Physics in University College Cork, and Louis Smith were secretaries to the group. The report appeared in December 1955, was accepted by the NFA in principle and sent to the Minister before Christmas, with a request for consultation. No consultation took place, though there was an informal meeting in June 1956.

The Research Institute Commission Report received praise from many quarters. The *Nationalist* said it 'was the most efficient report produced in the country for many years ... what a pity this commission had not been set up seven years ago'.[1] The main thrust of the recommendations was that an independent institute should work *through* existing institutions, where they existed and were suitable. The Department's research and advisory services, should be transferred to the Institute, giving a two-way flow of contact – bringing researchers into contact with the grass roots and placing the farmers, through their advisers, in close touch with the latest research. Agricultural colleges and the existing research stations should be transferred to the new institute.

These ideas, echoing the recommendations made by the Recess Committee and in Parliament before the Department was founded, could not expect a welcome from the Government. Dr Atkins, president of University College Cork, gave evidence of the Government's approach: 'When the proposal ... came before the Governing Body of UCC in March 1953 the Minister for Agriculture Walsh made it clear that in the event of UCC not agreeing to the transfer of Dairy Science the Government grant for its upkeep would cease.'[2] The Institute was to have the same independence as the universities. A hot debate developed round the country. A 'secret weapon' of the farmers was Dr Louis Smith, Agricultural Attaché at the US Embassy, who went far outside the normal bounds of diplomacy in promoting the ideas of the Joint Report at farmers' meetings. Whenever the Irish Government protested at newspaper reports of his speeches he could blame the other Dr Louis Smith, of the NFA.

The outcome was a substantial victory – though incomplete. The Foras Talúntais was as independent as is possible when receiving government funding. Dr Tom Walsh, its first director, was of independent mind and used the increased resources effectively in such areas as soil science and livestock breeding. Chairmen such as John Litton, Rory Murphy, Paddy O'Keeffe could hardly be faulted for lack of grassroots contact.

However, Juan Greene described the Institute as 'only a shadow of what we had hoped', and the representation of farm organisations as farcical.[3] The Governing Body gave little representation to producers. Of 35 members the Government would nominate the director and ten; third-level institutions 14. The ten representatives of farmers would be elected by 24 organisations having one vote each, with two votes for the NFA. In giving equal representation to the few hundred bee-keepers and the thousands of milk suppliers the apparent intention was to split the farmers and weaken their influence.

It was regretted by the NFA (at the 6 October 1956 National Council meeting) that the advisory services remained separate from the Institute, breaking the direct link from farm to researcher. In the 1970s we still find Paddy Lane resisting increased government control and asking for integration of advisory services with the research arm. The Government eventually took heed. Mark Clinton, Minister for Agriculture, integrated the three services in his National Agricultural Authority (1977). Next year the Authority was dispanded by the new Fianna Fáil administration. Ten years later another Fianna Fáil government again united research, education and advisory service in Teagasc.

The NFA regretted that the Government ignored other proposals of the Joint Report: That the Institute budget should cover a number of years to enable longer term planning; that the Institute should control the allocation of governmental research grants; that the Director be appointed for a five-year term and by the Council, not the Minister.

None the less, the position of research and advisory services improved under the Foras Taluntais. In 1991 Teagasc had 900 research workers in eight centres. There were 300 advisory staff. With the reduced number of farmers this gave a ratio of 1 to 700 holdings, an improvement on the 1955 ratio of 1 to 2,000. Some 30,000 farmers now pay for advice and a further 100,000 attend educational events. Specialist advisory services are integrated. Private advisory services have developed. The Irish Farm Accounts Co-operative, launched by the IFA in the presidency of Donal Cashman, and under his chairmanship, filled a gap.

The European Union lags far behind Japan and USA, with just over half their proportion of scientists and engineers. Within Europe Ireland's public expenditure on research is substantially below even Portugal and Greece. In line with the short-term view of Irish politicians, it was to be expected that research should be as neglected in agriculture as in industry. In agricultural research expenditure Ireland is lowest – half that of Belgium, a quarter of that in the Netherlands, one seventeenth of that in Britain. Even Portugal spends more. State support for university research is larger by half in Greece, the next lowest country.[4]

The foundation of An Foras Talúntais was crucial to the new association. It was the first public issue on which the NFA, with Macra, took a stand for a more open government policy and for independence in research and advice. By drawing on the knowledge of the academic and business interest, so long frustrated by bureaucracy, helpful contacts were made. A reputation for informed and constructive opinion was established. Though it was a heavy drain on finances, a fund of good will was built which was repaid in free expert advice.

INDUSTRIAL COMMITTEE

The Industrial Committee, under Tom Finegan of Louth, was handed every job that could not fit elsewhere – a 'rag-bag'. Each meeting new problems were presented with no common theme except their urgency.

Relations with Industry

The Association had to balance the immediate interest of the farmer with the need to favour and support industry in Ireland. Industrial workers were the food market and gave employment at home to the farmers' children. On gaining independence the Free State had the same level of industrial employment as India. Farmers supported infant industry in the 1930s by paying high prices. Some industries exploited their monopoly, continuing restrictive practices and out of date methods. The Association had to fight for reduced prices and increased efficiency in all sectors, while favouring industrial expansion.

Fertilisers

Fertiliser use, though increasing rapidly, was still half the optimum level recommended by the Research Institute.

The Industrial Committee was jolted into action when a new duty was imposed on imported fertilisers in June 1954. Macra (NFA was not yet born) discovered that farmers in other countries were paying less. This burden rested on the agricultural exporter, whose price was set by international competition. In 1955 the Association successfully opposed the manufacturers' application for a further price increase. A subsidy was given to bring the price of Irish manufactured product down to that of imports. The NFA was still dissatisfied. The Prices Advisory Body gave a forum to state their case in December 1956. The high cost was largely caused by expensive handling in Dublin docks and to the slow turnaround of ships. The farmer did not feel that he should pay for restrictive practices and avoidable inefficiency.

In 1958 an additional subsidy of £4 a ton (£1.75 million) a year was conceded by the Government. This subsidy was entered in governmental accounts as a subsidy to Agriculture though it only partly compensated farmers for the cost of protecting an inefficient industry. Particularly excessive charges were made for granulation and compounding of fertilisers. The price of compound was 23 per cent to 30 per cent dearer than that of the ingredients. In the case of some granulated compounds the margin reached 46 per cent. It was cheaper to mix with a shovel on a barn floor. The Irish charge for compounding was £2.75 a ton compared with 26 pence in Denmark. Some reductions were obtained. The battle continued for years. In November 1965 the NFA was still appealing to the Prices Advisory Body that compounders were adding 60 per cent to the raw material price.

Basic slag was tested and found to vary in quality. Rebates were obtained for purchases with under 40 per cent phosphate content.

In spite of complaints the Association established friendly liaison with firms such as Goulding's, Unidare and the chemical works, visiting the factories and discussing products. A mutual interest in success of farming and of Irish industry was recognised. A survey of prices showed a wide range through the country. Prices in the West were 25 per cent higher than in areas served by co-operatives. Charges for seasonal credit rose as high as 20 per cent p.a. The solution was found in publicity. Weekly reports of prices in all towns in the Irish Farmers' Journal opened nationwide competition. Co-operatives, such as North Connaught Farmers' developed to serve the 'co-operative desert' – the area with the highest prices.

Nitrogen Éireann

The first meeting of the NFA National Executive expressed concern at the prospect of a state-owned nitrogenous fertiliser factory. The project had a curious history. First proposed before the War as a means of stimulating demand for turf, the project lay dormant within Industry and Commerce for the duration. After the War it was revived, but relocated in Avoca to use waste pyrites from the copper mines as a source of sulphur. The mines closed but the project did not die. When Lord Wicklow left his country seat, Shelton Abbey, to the nation so that its beauty could be preserved, it was chosen for this chemical works. The river Avoca, being already polluted by the mines, supplied water. Sulphur emissions destroyed many trees. Eventually, the house became an elegant minimum security prison.

Dr Crotty, economic consultant to the ICMSA, reported to their National Council that the proposal was sound. Accepting the Department of Industry and Commerce estimates of a capital cost of £6 million and departmental cost projections, fertiliser should fall by about 10 per cent on current or foreseeable prices. He had, 'no difficulty in finding in favour of the proposition.'[5]

The NFA committee found that nitrogenous fertilisers were available abroad as industrial by-products or made where energy was cheap. An industry lacking such resources would find it difficult to be competitive. Farmers feared that they would have to carry another non-viable industry. The location was also questioned. It was some miles from the port through which materials and oil must be imported. Limestone had to be transported. The site was in a narrow strip of the East coast with mountains separating it from the chief tillage areas. Transport to tillage areas would be unnecessarily costly.

Instead of getting together to present a reasoned and agreed view based on research, the two organisations purporting to represent the farm interest disagreed without even knowing each others' opinion. The Government could only ignore such representation. There is no shame in reaching an honest opinion, but, in presenting contradictory submissions, the farmers were certain to be not only wrong but ridiculous.

On receipt of a written guarantee from Industry and Commerce that the price would never be allowed to exceed that of imported products, the Committee decided that it could not object further. Tom Finegan, Industrial Committee Chairman, was appointed to the board. The Committee Report of 1960 could claim that fertilisers were available at prices competitive with other European countries. The only remaining complaint was that the government subsidy, needed to bring home produced fertilisers down to import prices, was entered in statistics as a subsidy to farmers, rather than to the industry which was the actual beneficiary.

Sheltered from competition, NET accumulated losses of £117 million between 1974 and 1981.[6] By 1978 the debts of NET reached 800 per cent of its equity.

Agricultural Lime

Little lime had been applied to the land for a generation. After the War bulk road transport became cheap and limestone ground at the quarry replaced lime burnt in kilns. Subsidies encouraged the spreading of limestone directly from contractors' lorries. A beneficial scheme was blighted by the monopoly of the state-owned CIE. This company had not the truck capacity for the work and was uncompetitive in price. In 1955 it exacted a levy of £30,000 from the lime transport subsidy of £199,910 for its non-carriage of all tonnage distributed by private hauliers.[7] This charge was a matter of constant irritation to farmers and contractors for many years.

Agricultural Chemical Imports

Chemical manufacturing was protected in Ireland as part of the industrialisation drive. The Committee expressed concern on price and availability.

The small size of the Irish market caused problems. The Committee of Inquiry on the Chemical Industry[8] found that a plant of optimum size could supply the annual Irish demand for most products in about two weeks. The industry had to carry the cost of idle plant for the rest of the year. Plant became quickly obsolete as new, improved products were developed. There was a clear conflict of interest between the farmers, who demanded the cheapest and most up-to-date materials and the manufacturers, aided by the Department of Industry and Commerce, who restricted imports of improved products.

The Department of Industry and Commerce import licence required a 'letter of freedom' from every possible manufacturer to prove that equivalent Irish made supplies were unobtainable. The manufacturers saw no reason to answer such requests. Foreign tests were not accepted as evidence. Imports of the new chemicals were delayed till tests were repeated in Ireland. The Department took months to consider applications. The Minister's reply to parliamentary questions was that there 'was no undue delay'.

Machinery Duties

In August 1956 the Taoiseach, Eamon de Valera, called the NFA to a meeting. The two Irish machinery manufacturers were in a critical financial position. He proposed to allow them one year of tariff protection to make their production competitive. The NFA agreed, provided that the revenue from the protective duties was used for research and development. A joint committee of farmers[9] and manufacturers was set up to test Irish and imported equipment. The Department agreed to a review in a year. Pending the report of this committee the farmers had to pay the duty of 37.5 per cent full rate and 25 per cent on British imports. £20,000 was collected in the first year.

In breach of agreement with de Valera, levies remained on imports of machinery into 1958. No report came from the review committee – which was never called. The protected firms continued to refuse agencies to farmers' co-operatives.[10] There is no trace of the tariff revenue being devoted to research.

In 1961 the duties were still in force and the Department tried to extend import duties to other machinery. The Agricultural Institute, with which the NFA Committee co-operated in farm machinery testing, produced reports on various equipment but the Minister for Industry and Commerce, William Norton (Labour Party) of the Coalition Government, refused to include farm machinery in his review of protective duties. The Committee complained that this was a breach of the undertakings they had received.

Protective tariffs were not the only cause of high prices. Tractors from Britain cost up to £200 (30 per cent) more in the Republic than in Northern Ireland. This was the trading policy of the manufacturers. Price maintenance was enforced through agency agreements. The sale of a tractor from Northern Ireland to a Southern farmer was forbidden and would be traced when the machine came for servicing. The agent in the Republic reported his 'lost' sale to the manufacturer. The fine imposed on the Northern agent for breach of agreement was paid to the aggrieved agent in the Republic. It proved impossible to combat this practice. The differential tariff arranged under the Anglo-Irish Trade Agreements effectively excluded serious competition from continental manufacturers.

Removal of such practices was one attraction of the Common Market.

Compulsory Land Purchase

Power of compulsory purchase of farm land was given to local authorities and half a dozen state companies under various acts. Land could be acquired for almost any purpose within their statutory functions, even if there was no immediate plan for use in housing, water supply, drainage or road widening. These powers were strengthened in 1964 and 1966.

The individual farmer finding himself 'in the path of progress' was incapable of standing against the experience of the bureaucrat. Joint action, giving NFA a mandate to negotiate and hire engineers and valuers was arranged.

Way-Leaves

The Electricity Supply Board claimed the right to place its million poles and pylons at will and without compensation to the land owners, even where an alternative route could avoid the damage. One member, Reg Heathly, had 44 poles on his 35 acre holding – none of them serving his property. Furthermore, officials could enter the land for inspection at any time in spite of damage to crops. The position was felt to be oppressive and unconstitutional. NFA inquiries found that Ireland was quite unlike other countries. In Northern Ireland, for example, an annual payment of £0.5 million was made to farmers. Publicity and representations were without avail. John Litton received support from many sources, including a surprising handwritten note on Aras an Uachtaráin headed paper, saying merely 'keep up the good work'.

In 1968 compensation was obtained, amounting to £0.25 million a year and the ESB agreed to supply electricity to farms at the domestic rate.

Telegraph poles caused problems similar to those of the ESB, though less damaging. Stratagems such as keeping a bull loose in the threatened field were ineffective. Some farmers moved the pole markers to more convenient places as soon as the surveyors had gone. Neville Chance was more successful. He found a selective weed killer carrying a warning notice with a red skull and cross-bones which he showed to the workers' union representative. It threatened death and sterility if the yellow stain showed on workmen's trousers. Repeated spraying diverted the telephone line round the headlands of a valuable field.

A 'Wayleave/ Land Acquisition Manual' was prepared.[11]

Water Supplies

Excess water is the usual Irish problem but domestic water supplies were always inadequate. In the Census of 1946 there were 307,000 farm dwellings, of which 17,000 had piped water. Half of those had baths. Surface water, on which most farmers relied, could be contaminated. The woman of the house had the job of carrying water to the house. No health cure, especially when a baby was expected, it shortened the life of many a farmer's wife.

Rural electrification and plastic piping made local co-operative water schemes possible without calling on expensive skilled labour.

A conflict arose with the public authorities. Engineers and county councillors saw regional water schemes as a means of giving employment: rural people wanted a cheap service. Extending regional schemes to the scattered rural population cost £300 a house – £30 million for the whole country. With interest and depreciation this gave an annual charge per house greater than a month's

15 Domestic piped water relieved women in particular.

agricultural wage. The co-operative schemes cost the state a maximum of £150 and local volunteer labour was unpaid.

NFA branches complained that complexity of the various government grants impeded the work and that discrimination in favour of elaborate regional schemes placed an unnecessary burden on local taxation. The cost of this essential service was inflated by protection of industry. The duty on plastic piping added 10 per cent to the capital cost and electric pumps were also subject to duty.

A running battle was fought at local and national level to solve the old problem: should the State levy taxes so that people will have what the State thinks good for them; or is it better to let them keep their money and provide the service they want for themselves?

Mining

Open cast mining operations and oil and gas pipelines began to encroach on farm lands. When disputes on compensation and methods of restoring land arose information was gathered from farm unions more experienced in handling this kind of dispute. The Minister for Transport and Power gave a sympathetic hearing (1962), though the legislation was a few years in enactment.

Building Materials

Farm improvement grants were conditional on the use of Irish materials. In principle farmers approved of this measure of industrial protection. In practice they complained of rapidly rusting corrugated iron roofs and barbed wire. The manufacturers replied that their zinc coating was thicker than that on imports. the NFA approached the Institute of Research and Standards, which traced the easily remedied faults to soft core material in the wire. This stretched under tension and cracked the zinc coating. The corrugated iron had pin-holes in the zinc giving rust a foothold, a matter easily rectified.

Drainage

Drainage, arterial and on farm, had been long neglected. John B. Keane, the playwright, wrote in the NFA Year Book of 1966 of the Kerry small farmer who boasted that a man on horseback could not cover the whole of his farm in one day. 'Tis true for me. Horse and man would be drowned after the first field.'

Major floods required organisation of voluntary help giving fodder to flooded farms from the surplus of drier areas. Co-ordination was needed in clearing outfalls for water for the benefit of neighbours up-stream – though the Committee asked, unsuccessfully, for compulsory powers.

Land reclamation and drainage were grant aided under many schemes. The NFA, acting through this Committee, helped the Government to make the system user-friendly in design, credit facilities and index-linking of grants.

Tolls

Native government had not tidied up legislation dating from the days of ancient monarchs. Tolls were rights given by royal charter to individual favourites and their heirs, or to towns such as Bandon. Charges were levied on all sales in the area. Cattle marts and wool auctions were bound to pay 'for the non-use of the landlord's scales' even though these sales were on the co-operative's premises and the lord provided no service whatever.

A hundred years earlier British government commissions had recommended that these monopoly charges be abolished.[12] Independence has not led to reform. Though many toll charges have fallen into disuse, Carrickmacross, Co. Monaghan, Urban District council paid £55,000 in 1993 to buy out the private owner of rights granted by Queen Elizabeth in 1576.[13]

Transport

Farmers sell bulky, perishable produce from scattered holdings to distant markets. Transport charges on such low-priced goods can be crucial.

The story of modern transport regulation begins with the building of the railways. To protect the public from the rail monopoly the Government regulated working conditions and charges. When competition from road haulage threatened the existence of railways after World War 1, the Railways Road Services Act of 1927 permitted railway companies to buy lorries. Conservative management wasted their opportunity, treating lorries as mere feeders to the railway.

The Road Transport Acts of 1932 and 1933 attempted to make railways viable by extending their monopoly to the roads, an attitude haunting all future Irish transport policy. From an agricultural point of view the situation was described as 'A Tragic Comedy in Three Transport Acts'. Few licences permitting road haulage for hire were issued, and those only to lorry owners already in the business. Licences could be transferred only to members of the same family. The railway had power to buy up and suppress licences. By 1938 Road Passenger licences had been reduced from 107 to 48 and merchandise 'plates' from 1,356 to 886. Cheap transport of people and goods became secondary to the maintenance of vested interests in the pre-existing structure.

The Great Southern Railway petitioned the Transport Tribunal of 1939 that private hauliers be limited to 15 miles from home. The Transport Tribunal rejected this limitation on the citizens. The Majority Report gently said that 'the company had shown no aptitude for road transport'. The Minority Report, more pointedly, drew attention to the fact that GSR costs were 11 pence a mile compared with 8 pence in a private company.

Firms found it cheaper and more reliable to buy their own lorries, though it meant under-utilisation of equipment and empty return journeys. Privately owned lorries formed 94 per cent of the national fleet in the 1950s. The Minister considered that every van and horse drawn cart was a competitor with the public haulier.[14]

The citizens' complaint was not only on price. Service was described as unsatisfactory and subject to interruption by strikes. There is an old saying that 'the first dividend of a monopoly is a quiet life'.

Transport Tribunal 1952

CIE petitioned that privately owned lorries – 94 per cent of those in the state – should be limited to a radius of 20 miles from base. Furthermore, state companies should use CIE exclusively. The ESB, for example, should not own any lorries over two tons.

The co-operatives, through the IAOS, argued the farmers' case in detail. Farmers were often criticised for spending long hours delivering a few gallons of milk to the creamery. The price of the few gallons of milk might not cover an agricultural wage for the delivery time but cash was desperately scarce, while horse and man were underemployed. When co-operatives arranged collection by contractors with horse-drawn 'long cars' at a charge of a penny a gallon, hard cash was so scarce that many small farmers refused the service.

Because CIE considered the haulage of milk unattractive, special exemptions were given yearly to unlicensed hauliers. In the Northern and Western areas the average herd was about four dairy cows, sending under 700 gallons a year to the creamery. It took 39 farms to provide full load for a tractor.

Till 1952 these milk haulage contractors had paid road tax at the agricultural rate (£5). The Finance Act of 1952 raised the tax to that on a commercial lorry, £31.50, and removed the fuel tax rebate. The attitude in the Department of Industry and Commerce was illustrated when an official commented to a deputation that CIE had shown restraint. They did not prosecute hauliers who carried the odd pound of butter back to the farm.[15]

NFA estimated that the tax became about 1 penny a mile on a low-mileage tractor, compared with 0.15 pence for a small lorry and 0.1 pence for a car. The ass and cart, once so common at the creamery, had disappeared but the farmer who could not afford a horse could afford a tractor even less. Some 4,000 farmers were affected. The typical case was a widow whose couple of gallons a day was carried by a neighbour making his own daily delivery. The tax was prohibitive when the price of milk was 5 pence a gallon. Where a haulier was obliging a few neighbours his revenue was less than the extra tax.[16]

Till the 1950s the farmer met his local needs by horse and cart. It never occurred to anyone that this ancient right should be restrained. Between 1947 and 1955 the number of horses used in agriculture fell from 350,000 to 250,000, while the number of tractors licensed for the road increased from 2,976 to 14,819. By 1963 horses had fallen to 136,800 and tractors risen to 40,488. The new tarmacadam roads were unsuited to horses. The transport revolution was ill received by officialdom. Instead of recognising that the tractor had replaced the horse they spancelled it with restrictions and tethered it to the shortest of journeys.

The NFA was immediately concerned with the chaotic legislation. For the government Committee of Inquiry into Internal Transport (of which Dr Juan Greene was a member) the submissions already made by the IAOS were developed further.

The farmers' complaint was:

- The tractor was not allowed to do the normal work of the disappearing horse. Consequently, farmers were left without their essential local transport. Small farmers, unable to buy their own tractors, were penalised if they co-operated to provide their own transport.
- Town depots could not serve rural needs efficiently. The drivers lacked local knowledge and incurred additional mileage costs in travelling empty from town depots to collect cattle.
- Labour in CIE was undependable because of strikes and expensive. CIE drivers received twice the agricultural wage and more over-time.
- Being in direct competition with farmers abroad, Irish farmers should not be forced to bear heavier cost in road tax or unnecessary restraints than their competitors.

The Committee of Inquiry found:

> We have been impressed by the view that much of a farmer's transport requirements – particularly the short haul of small loads which must often be done in conjunction with other farming operations – is of a nature which cannot be conveniently or economically catered for by public transport. We have been impressed by the view that it is in the national interest that the maximum degree of co-operation between neighbouring farmers should be facilitated and that a tractor owning farmer should not only be able to undertake his neighbour's tillage and harvesting etc. for payment but that he could be permitted to undertake the local transport needs of his neighbour on a like basis ...
>
> It is desirable that the many farmers who have not been or may no longer be within the range of horse transport of their nearest railhead should be able to avail themselves of their neighbour's transport for the collection and delivery of their goods at the nearest railway station ...
>
> Accordingly we recommend that the provisions of the Road Transport Acts should be amended to permit farmers to carry for reward by means of tractor and trailer the goods of neighbouring farmers resident within a radius of two miles of the tractor owner and that the carriage for reward of the farm produce and farming requisites of neighbouring farmers within this radius should not affect the eligibility of the tractor owner for the present taxation rate of £8.[17]

This seemed to give all that the Transport Committee of the NFA had demanded. In effect they asked that the tractor be recognised as the replacement of the horse. There was no need for a limit to distance. Tractor haulage could not compete with trucks on the long haul because it was too slow and the tyres wore too quickly.

The Transport Act of 1958 conceded only minimal relaxation. Farmers could carry cattle for neighbours living within two miles to a mart within 20 miles, only on the day of a sale. A similar concession was given for the haulage of wheat in harvest time. In fact the position grew worse and restrictive measures multiplied.

CIE was at the root of the most serious problems. A transport monopoly should, in theory, give economies of scale and provide a social service to small customers and scattered populations. When it failed on all counts the response was to increase subsidy and impose stricter restraint on competitors. The objective was the preservation of CIE rather than the service of farming.

In many cases CIE could not provide the service.

Approached by NFA, CIE expressed their anxiety to help farmers by providing haulage licence 'plates' provided:

1) Only agricultural produce was involved.

2) Special need was shown.

3) No interest of CIE should be affected.

4) CIE should have sole discretion when issuing the plates to impose limitations as to –

a) traffic to be carried;

b) area of operation;

c) duration.

5) Information on goods carried etc. to be given to CIE

You will appreciate also that, in order to maintain control, the arrangement would have to be confined to farmers who are sole owner ... to one address'.[18]

Few, if any, 'plates' were forthcoming.

Haulage of Produce

With the demise of the fair, farmers took their cattle to marts. The catchment area of a mart was some 20 miles, far beyond walking distance. Lorries from the central depots of CIE were unable to give satisfactory service. The farmer had to pay for the empty journey from the depot to the farm and back to the depot. In West Donegal the CIE lorry had to travel 82 miles empty, from the depot and back, to carry cattle 16 miles to the fair. A local tractor or haulier could have avoided this cost.

The average farmer did not have a full load of 20 cattle for a lorry. A local haulier could combine loads. The driver from the town had trouble and delay in finding isolated farms at 6 o'clock of a winter's morning. Cattle might not be sold and require transport home. Waiting time for a CIE lorry was 55 pence an hour, compared with nil for the tractor.

Tractor Taxation

Aided by the Agricultural Machinery Traders' Association, the NFA found the level of tractor tax in other countries to be lower than in Ireland and the restraints less (see Table 7).

The farmer in the Republic was at a further disadvantage in driving license and insurance. These cost £13.50 a year in the Republic compared with £5 in Northern Ireland.

The Minister for Industry and Commerce, William Norton (Labour), called the £8 tax a 'subsidy' and said that the farmers were no worse off than before – there were plenty of horses.[19] An earlier deputation to the Department of Industry and Commerce met an even colder response. NFA representations were successful. In 1964 the basic rate was reduced to £2.50, equal to the Northern Irish rate but still above other countries. The milk haulage tax was removed in 1956; the

Table 7 Taxation and Conditions of Tractor Use in Certain Countries, *c.*1957

Country	*Yearly Tax*	*Comment*
United Kingdom	£2.00	A further tax of 25 p. entitled the farmer to haul any goods for farmers living within 1 mile.
France	£1.50	A local tax.
Germany	zero	Reduced petrol tax.
Netherlands	zero	
Sweden	zero	Petrol tax refund of £23 p.a.
New Zealand	zero	If used within round trip of 13 miles. Fuel tax rebate. Third party insurance £0.5 pa.
Ireland	£0.25	Use with farm instruments only.
	£8.00	Haulage of own farm produce on roads.
	31.50	Farm haulage contractor, co-op or farmer with second occupation.

co-operative tax was reduced to £2.50 in the 1966 Annual Review. Insurance charges were controlled by competition when the IFA founded the co-operative FBD.

Restraints on Tractor Use

Faced with absurd legislation the lawyers played word-games to catch the new criminals or evade penalties. What was a 'tractor'? A farmer was fined £108 when his tractor and trailer was held to be an 'articulated vehicle'; another was summoned because a platform on the back of the tractor made it a 'lorry'. Farm tractors could not carry 'merchandise', but that too was hard to define. Cattle brought to the fair were adjudged to be 'merchandise', whose haulage was restricted. Sheep, on the other hand were 'agricultural produce' and so exempt, but West of Ireland farmers complained that they were hindered from bringing pigs to market. A farmer was prosecuted for carrying logs. Though produced on his farm these were not 'farm produce'. A Skibbereen farmer was before the court for carrying home a table and chairs with his tractor from an auction. They were for use in the farm house, but not on the farm. Farmers were fined because workers took a ride to work in the trailer.

At a court in Kilcormac, county Offaly, a farmer was charged that he used his tractor to carry water for his stock. The Garda Sergeant further submitted that when the tar barrel was empty it was neither agricultural produce nor a farm implement. The case was dismissed with laughter in the court (4 February 1956).

Bandon cattle mart was reported as 'beset with law enforcing gardaí'. Bandon is 19 miles from Cork City. The zone of freedom for haulage only extended 15 miles from the city centre. Cattle coming from the 'free' zone had to be unloaded from their trucks and walked the last four miles to Bandon mart.

Small farms were particularly penalised. Often the bohereen was unsuited to a lorry. Potatoes had to be brought from a field where a lorry could not go. A tractor was the obvious solution. The neighbouring tractor owner was willing, but forbidden to oblige.

At the end of the war horse numbers were reduced and any tractors were at least 10 years old. There was a danger that the few machines coming on the market would go to the rich. If a subsidy was given to the small farmer equipment would be under-used. In other countries a policy decision was made to make the best use of scarce supplies by encouraging co-operative use of machinery. In France there were about 8,000 such co-operatives; in the Netherlands 500. The system was impossible in Ireland. Farmers ceased to be treated as farmers if they co-operated to own a tractor. Common use, even for members' produce, was forbidden. Besides that a co-operative had to pay four times the road tax of the individual. The result was a wasteful over investment in tractors in Ireland.

Any farmers with a second occupation, as shopkeepers or doctors, were subjected to the higher road tax of £31.50 even if the tractor was only used on their own farms. The penal tax on tractors owned by co-operatives was removed in March 1966, but balanced by an increase in car tax.

Farmers were joint owners of creamery lorries through the co-operatives, but the carriage of the individual farmers' goods or produce was forbidden.

There were numerous lesser annoyances: The 25 pence tax for tractors only permitted passage of the tractor and its implements on the road. In principle this did not permit a farmer to carry his own seed and fertiliser across the road from one gate to another. Ownership extends half way across the road from the hedge, so in law the farmer never left his own property.

In 1964 lighting regulations were introduced for all road vehicles. Lights had not been standard makers' equipment on tractors. Instead of penalising the offence of driving without lights the law proposed that all tractors be fully equipped, though in most cases they were never on the road by night. The necessary modifications would have cost £1.5 million. An exemption was negotiated for existing machines.

'Scut-Trucks'

For years farmers made representations to free the regulations on farm trucks. Jeep-type vehicles, like the 'pickup trucks' so common in America, could be adapted to various loads and farm work, especially if fitted with a four-wheel drive. They could save the capital cost of a car for social use.

A Kildare farmer was fined £418.50 (mitigated to £50) for drawing his own cattle to market (1967) in a trailer behind a cut-down lorry. An appeal was successful on the grounds that no weight from the trailer rested on the vehicle; therefore the cattle were not 'on' the lorry. This established the legal position that 'scut trucks' were within the law, provided that there was no passenger seat and the trailer hitch was behind the rear axle. A Tipperary farmer established that such a truck could use untaxed diesel fuel.

The NFA won a case for liberalising use of these vehicles in the Circuit Court; lost it in the High Court; lost again in the Supreme Court. Tax was further raised to £50 in 1973 and they were forbidden the use of untaxed 'red' diesel.[20]

Wexford fruit growers had a variation on the same theme. During the harvest they had small quantities of soft fruit for urgent delivery. When they used their private cars prosecution followed. The wisdom of the court decreed that if the seats were removed there was a basic structural alteration in the car and it became a commercial truck. In future raspberries had to travel on cushions.

The Gardaí are not by nature vindictive upholders of stupid restrictions. The pressure was placed on them and on the Department of Industry and Commerce from CIE and the trade unions for the protection of the monopoly.

The farmers were not opposed to regulation for safety. For example, the request from some branches that youngsters of 15 be allowed drivers' licences was rejected by the NFA Transport Committee. There was no request for tax exemption or for rights not available to the town businessmen or to competitors overseas.

Brian Lenihan, Minister for Transport, announced legislation removing the carriage of cattle, sheep and pigs from the scope of the Road Transport Acts, so as to allow their carriage for reward without licence (24 October 1969).

'Putting an Ass in Front of a Tractor'

In 1956 the British and French invaded Suez, causing an acute fuel shortage. The Department of Industry and Commerce introduced regulations to conserve supplies of fuel without consulting any farm organisation. Designed for urban conditions the result was a comedy.

CIE was given full supplies. The Minister for Industry and Commerce offered CIE any further supplies they might need (18 December 1955). Tillage farmers were allocated a monthly ration of petrol or diesel for five days' ploughing – this at the hight of the ploughing when the Minister for Agriculture had set the target of an extra 100,000 acres of barley. In Britain farmers were given priority, with 90 per cent of normal supplies. In Ireland TVO was unrationed, but only enough petrol allocated for stopping and starting the tractor 12 times. Meanwhile a ration of petrol was given for pleasure motoring. The farmer was treated as equal to the suburban traveller who had a bus at his door. In Britain farmers' longer journeys were recognised with a treble car ration.

To limit consumption fuel could only be supplied directly to a vehicle from the pumps. No sales could be made on Saturday or Sundays. Every farmer with a car or tractor had to drive to town for fuel, but not on the day he would be there for Mass.

Such mishandling was best combated by ridicule in an article, 'On Putting an Ass in Front of a Tractor'.[21]

Labour Relations

A crucial problem was that the lorry driver was paid urban union rates and overtime – three times the agricultural wage of a tractor driver or the small farmer's income.

Apart from the wage differential between rural labour and the unionised urban labour of CIE a cultural difference caused friction. Kindly farmers who gave the CIE drivers a meal when they came in late were incensed to be charged for the delay – at overtime rates. There was a clash of cultures between the trade unionist who sold his time and the farmer who did not price his hours of work and considered his hospitality had been abused.

Another union practice to which attention was paid was the 'employment' of helpers on lorries delivering grain whose only function was to keep the driver company. In practice the unemployed would wait round the corner from the grain mills. The lorry owner paid them to sit in the cab during unloading to fulfil the regulations.

The Docks

Export depends on efficient work in the docks. The producer can only get the market price abroad minus the costs of transport Therefore abuses in Dublin Docks concerned the NFA.

The cost of transport across the 125 miles of the Irish Sea to Liverpool was greater than that of meat from the Argentine 6,400 miles away. This had been submitted in evidence even to the Agricultural Commission of 1923. In economic terms an Irish meat factory was further from its London market than one in Buenos Aires. The cost of Cross-Channel transport was noted as 'the most serious disadvantage to Northern Irish industry relative to that in Great Britain'.[22]

One cause alleged by Carter and again by Desmond Branigan, Secretary of the Seamen's Union,[23] was the development of monopoly by two firms, Irish Coast Lines and British Railways, since 1938. In 1926 the Irish companies, B&I and the associated Cork Steam Packet company had 50 ships. In 1957 they had eight. In 1958 even B&I was owned by Coast Lines, though their ships were registered in Ireland. Dissatisfaction with the service was frequently expressed but the response of the British Railways representatives at a meeting attended by the NFA in the Department of Agriculture did not inspire confidence.

The charge for transporting live cattle was lower than for beef, though 20 ton of beef in a container is more compact and requires less handling than ,say, 40 live cattle. This, added to the price advantage given to stores fed in Britain under the Trade Agreements, was a significant deterrent to development of slaughtering in Ireland. (The Commission of 1923 and that of 1900 also observed this bias.) While no evidence could be produced to show government influence, it was remarked that the pricing structure followed British government policy.

Other problems were clearly of Irish origin. For example, in the export of turkeys mentioned in Chapter 5.

'Deep-Sea Drovers'

When beef exports to the Continent opened up the importing countries, like the British, wished to retain the employment and added value of processing in their own countries. They imported live animals, which had to be attended on the voyage. This became the job of the 'deep-sea drovers', a privileged group of 'button men' who had priority in hiring. The right of membership was hereditary.

Due to the riotous behaviour of these men, the police in Rotterdam refused entry to ships carrying them. Seamen had to be hired for the work, while the 'ghost drovers' stayed home. However, the drovers continued to draw full pay of £14.70 per trip while the ship was at sea, with additional pay for their discomfort in stormy weather and for over-time. An agricultural wage (£4.75 a week) was one third the reward for idleness.

To give the union officers their due, they offered the NFA sight of their letter sent to Sean Lemass, Minister for Industry and Commerce, at the start of the practice. It asked him to take steps to stop the abuse before it became established. The union assured him that they would not resist, but the Minister had failed to respond.

A German trade delegation gave this surcharge on transport as one obstacle to development of cattle exports.

Import prices were also inflated by unnecessary costs at the ports. Workers refused to use unloading equipment already in place. In Limerick grain was unloaded from the ships to horse carts and carried across the road to be unloaded to the silo. The over-head transporter, direct from ship to silo was 'blacked' by the workers.

RATES/TAXATION COMMITTEE

Taxation was a rather minor concern in the immediate post-war period. In 1948 Central Government took 20 per cent of National Income and Local Government three per cent. Income Tax was five per cent of National Income, with only 6,000 people liable at the higher rate compared with 40 per cent of earners to-day. Few farmers and no farm workers reached the Income Tax threshold. Insurance was paid by one simple stamp.

As the Welfare State developed government expenditure took a growing share of National Income, rising to 40 per cent in 1965 and 67 per cent in 1981.[24] There is no such thing as a free lunch. When the money had to be raised by taxes and (increasingly) by borrowing, acrimonious debate was inevitable. Tax is mentioned in the minutes of almost every NFA Council meeting over long periods.

When farmers were pilloried as paying less than their share, it was the NFA's task to reply.

Much of the trouble arose from the haphazard development of the tax structure over many years, with confused terminology and changing importance of various taxes. The confusion took many forms:

Rates

The main farmer's tax, were first imposed to pay for local services. They did not appear in the Central Government's budget and so escaped debate in the Dáil. By throwing costs, such as health services, on to the Rates the Government concealed tax increases and avoided unpopularity. The impression was given that farmers paid no tax.

The county manager, under the 'reserved functions' of the Managerial Act (1942), ran the county. The money had to be found from local taxes, principally Rates. Should the County Council fail to agree the rate, the Council was suspended and the manager took over. Debate on income or expenditure was equally futile. If the revenue from rates was insufficient to cover the cost of services the money had to be borrowed. Already in 1959 debt service took 59 per cent of rates in Dublin, 25 per cent in Donegal, Cork and Carlow. The Irish State had already become addicted to borrowing as 'the opium of the people'.

Rates were payable on the poor law valuation of land and buildings made by Griffith about 1856. Land values were never revised, so farmers paid tax on the basis of their great-grandfathers' rents. No allowance was made for changes in markets or crop prices; in costs, method of production or family commitments. No sector of the community could be content with such a system. In a time of rising taxes and declining income, such as 1956 and 1966, discontent became acute.

The burden of Social Welfare increases with poverty. Accordingly the local charge was heaviest in poor counties. The tax was in inverse proportion to ability to pay.

Health services were a good example. Needed improvements were charged on the rates. Eamon de Valera, as Taoiseach, guaranteed that the cost, then 1.8 pence per £ PLV, would never exceed 2.5 pence. By 1966 health had become 35 per cent of the total local expenditure. The lowest county charge for health was 55 pence per £PLV in Dublin and reached £1.96. in Donegal (an increase of 60 pence on the previous year). Improved services and salaries were debated by the Dáil, which did not have to collect the tax. The counties had no power to change the level of pay or service.

Income Tax

Income tax took off on an upward spiral, made easy by inflation. Nominal tax rates rose little, or were even reduced, but ever more people, with ever lower real incomes, were caught in the net. The number in PAYE rose from 220,000 in 1961 to 740,000 in 1973. Eventually, due to inadequate adjustment for inflation, 40 per cent of industrial workers became subject to the higher rates of tax intended for the rich. In the 1960s farm workers' wages crossed the tax-free threshold of £300 a year, giving a liability of 12 pence. Concerned at the formalities involved, the NFA approached the Revenue Commissioners. They were assured that they could ignore the matter. Income Tax would never be imposed on those with minimum wages. In the light of future increases of direct taxes on the poor the NFA was wrong to accept this assurance.

Farmers' income tax, being assessed on a notional multiplier of PLV, did not immediately suffer this inflation effect. Naturally towns people were upset and hunted those not bearing their share of misfortune.

Tax Commissions

The NFA submission to the Income Tax Commission of 1961 used national income statistics and the farm management surveys of the Agricultural Institute to show that the great majority of farmers had incomes below the tax level; that some large farmers were paying up to 30 per cent of their income in Rates. After submission of the NFA case on taxation the official national income figures of farm income were revised – due to NFA representations. The Commission Report then criticised NFA for using inaccurate figures. The Economic and Social Research Institute report bore out the case made by NFA.[25]

The issues were debated with Garret Fitzgerald, later Taoiseach, in the *Irish Times*.[26] FitzGerald expressed a widely held urban view that, 'one seventh of the revenue from taxation goes to a section of the community that pays no direct taxation whatever'.

The NFA defence was:

1) FitzGerald considered only Central Government tax. Farmers paid £22 million in Rates, in addition to some Income Tax under Schedules A,B and D.

2) Rates, being assessed on the same presumed income as in these schedules, were in fact a form of income tax.

3) Rates are payment for general government services (such as health and social services) not for local services given to farmers.

4) The estimate of 'subsidies to farmers' given by the Minister for Finance, Dr Ryan, was padded out with:

a) Subsidies to fertiliser manufacturers to bring their costs more in line with imports, costing £350,000

Rates versus Income Tax

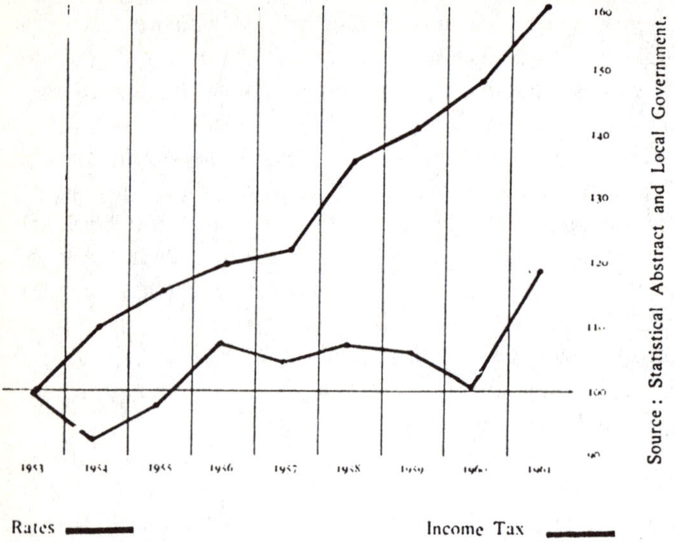

Cost of Health and Mental Hospitals per £1 p.l.v.

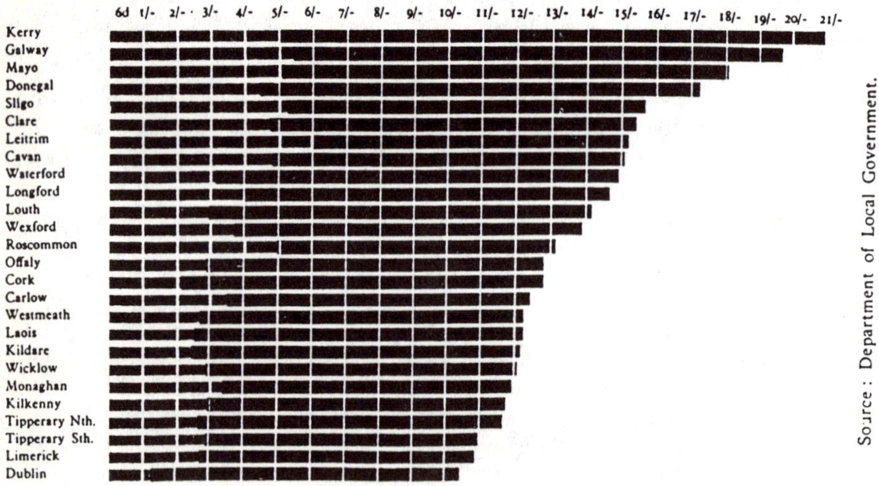

16 Graph A: rates versus income tax; Graph B: cost of health and mental hospitals per £100 PLV; Graph C: net rates payable on holding of £100 PLV; Graph D: net rates payable on holding of £20 PLV

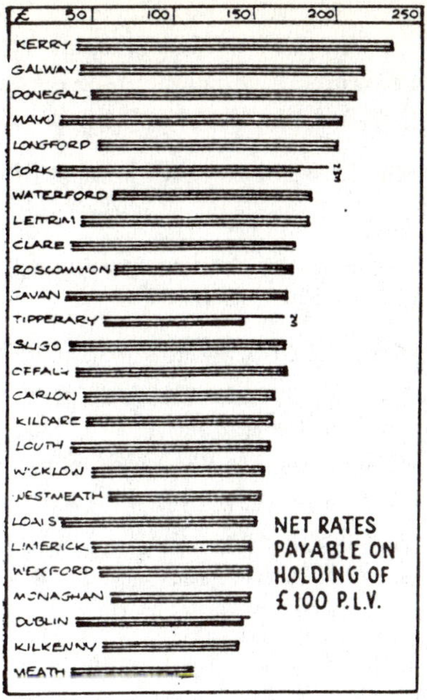

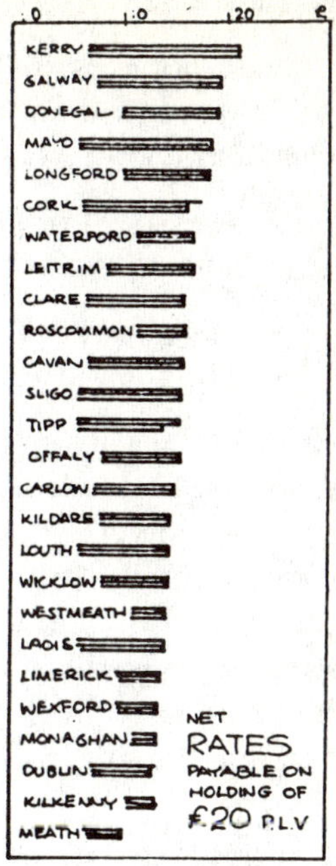

b) Grants to bacon factories, £30,000

c) American Grant Counterpart Funds, not paid from Irish taxation.

d) Agricultural grant, a partial relief of rates on the first £20 Poor Law Valuation of farms. If tax relief on low incomes was a subsidy then the 'subsidy' of earned income relief to income tax payers was £40 million.

e) Halving of land annuities, an old issue of the Economic War. If this was to be reckoned a subsidy, then the non payment of interest on loans to state companies must be also a subsidy to industry.

f) Payments to agricultural colleges and even to the Botanical Gardens (a park in a Dublin suburb) were reckoned a subsidy to Agriculture, but payments to vocational schools or professional university faculties were education.

NFA demands were:

1) Tax should be assessed on the average of several year's income, as in Australia. Taxpayer with variable incomes (like farmers) are caught in the high tax bracket in a good year but are unable to claim full allowances in the bad.

2) Notional assessment of farmers' income was favoured to save the cost of account keeping.

3) Costs, including earned income relief on unpaid work of wives and family labour, should be allowed against tax, as for any other business.

4) There should be a uniform national tax. Prosperous areas had few poor to be assisted, so their tax burden was light; poor counties placed heavy taxes on the their few large farms. The rich lands of Meath had half the tax rate of Mayo (see table from Year Book 1959). The Leitrim County Council told the NFA Rates Committee that it was the county councillors' duty in local patriotism to push the Rates as high as possible. There were only 40 major rate-payers in the county, the rest of the money flowed in from the Exchequer. It was necessary that 40 men be sacrificed for the good of the county.

5) Export industries such as agriculture could not bear taxes substantially in excess of those imposed on its competitors. NFA considered that their presentation of the tax systems of farmers in other countries was relevant.

6) Profit from export of industrial goods was relieved of tax. It should be given on equal terms to stimulate agriculture, the main export industry.

7) The Griffith valuation, was over a century out of date. Major changes occurred. Land near towns, like the poor sandy soils of North Dublin, had become valuable; heavy soils, such as the Maccamores in Wexford, had become marginal; land now under the sea due to erosion could not be derated; Monaghan land, profitable for flax growing in 1860, held its high valuation while Meath's valuation still reflected the poor cattle trade of that period.

The Government, estimating that a full revaluation would cost £10 million, would not undertake revision. The results would have been painful to many farmers.

8) Individual peculiarities were common. Absentee landlords requested a high valuation to impress London society; land agitators [like the Brangans' of North Dublin] were punished with high valuation; one half of Bray Head was valued £1 an acre, the other at a few shillings – because Lord Meath could appreciate the view but his small farm neighbour could not. Some farmers had their buildings revalued several times; others were never visited by valuers. Those well advised enough to leave one sheet of corrugated iron off the roof of a hay-shed escaped. 'Unfinished' buildings could not be valued.

9) Rates took no account of ability to pay caused by personal loss, harvest difficulties or bad prices. Between 1946/7 and 1958/9 Rates increased three times as fast as farm prices and six times as fast as farm family incomes.

The NFA action included: Request for examination of each county's operations by efficiency experts (1960). Repeated protests against the appointment of Rates Collectors. (These appointments were jealously guarded by the political parties as rewards for political service or for nepotism. It was held in court that not even a primary certificate was necessary as a qualification.)

In 1962 protest marches took place in every county, followed by a refusal to pay the first moiety of rates. Farmers were jailed, but released and a settlement reached. Negotiations led to derating of first £20 PLV on small farms, freeing 70 per cent of farms from payment.

Death duties, against which the NFA protested, were abolished.

Arguments that the tax on farmers trading through a co-operative was a tax on the individuals were unsuccessful.

The NFA arguments were vindicated and rates abolished when the farmers of Wexford obtained judgement in the Supreme Court that rates were unconstitutional and unjust (see Chapter 13).

Concession of an Employment Allowance against Rates was obtained in 1967.

CAPITAL AND CREDIT

Farmers live poor and die rich. In land and stock they require a greater investment than heavy industry, but the annual income is less than that of an industrial worker.

During World War 2 material for investment was simply not available. Worse still, compulsory tillage had drained fertility when total fertiliser imports were only £25,000 in 1942. Stocking was low.

If the farmer was to escape poverty by increasing output an extraordinary reconstruction effort was needed. The investment required was far beyond the farmers' own resources. Borrowing for production was essential.

Government policy and expert opinion was expressed by the Committee of Inquiry on Post Emergency Agriculture.[27] Credit did not merit a heading in the Majority Report. It was virtually dismissed in a two paragraphs (s. 18 & 302) concluding that farmers used little capital; did not plough back their savings; were adequately served by the existing system. The Majority Report judged that Irish agriculture was going nowhere and that export markets would close. On that basis the conclusions were defensible. Only the Minority Report of Dr Henry Kennedy, secretary of the IAOS, made a realistic assessment of capital requirements.

The NFA could not adopt so negative an attitude. Heavy investment was needed to bring fertility and equipment back even to low pre-war standards. Farmers were now solvent. Even the 5,600 farmers with debts deemed irrecoverable in 1937 had cleared their accounts. The debt of land annuities had been halved in the 1930's and halved again by the fall in the value of money. However, wartime savings, eroded by the falling value of money, were no longer enough to restore depreciated equipment. In effect the farmers had been paid in fool's gold which withered away by the time they came to spend it.

It was not enough to restore the position of 1938, the time of depression and the Economic War. To fulfil the government targets of the Programme for Economic Expansion the NFA calculated capital needs for the six years 1964 to 1970 at £308 million.[28] Farm income would increase by 36 per cent at constant prices. Such funds, needed mainly in the early years of an expansion programme, could not be saved from an agricultural income of £110m. a year.

Surveys of Credit

The Credit Committee of the NFA joined forces with Macra, the Irish Countrywomen's Association and the Agricultural Officers' Association to suggest action for improvement. The first requirement was information.

A first recommendation of the NFA National Credit Committee was that a USA Technical Assistance Project should prepare a full report on farm credit in Ireland. This was done by Fred Gilmore, Deputy Governor of the Farm Credit Administration, USA.[29] The Farm Management Surveys of the Research Institute detailing farm income, were supplemented by the Committee's own survey of opinion from 118 branches and 84 Agricultural Instructors[30] and comparisons with financial costs in EEC countries.[31] Government reports were available.

Irish farmers were often paying several times the interest charged on the Continent. Often the variation was due to the method of subsidising credit. Interest subsidies were the preferred means of the Continentals, reducing Italian 30-year loans to one per cent per annum in some cases. Tractor purchase loans cost as little as three per cent. Ireland relied on capital grants for specified investment. This was costly to the Government in two ways. 1) In time of inflation a grant is paid in full at the beginning of the transaction. An interest subsidy is paid later when money has fallen in value. 2) Administration of an interest subsidy is simpler and less costly. The lender, rather than the government, assesses the project. If a loan fails the interest payment ceases and with it the government subsidy. If a grant aided scheme fails the grant money has been lost in full.

The most serious faults in Irish credit schemes lay in the absence of loans or grants for working capital.

Sources of Credit

Instead of an active market of saving deposits and loans of long and short duration, the farmer saved money for the rainy day under the bed or in bank deposits paying minimal interest. These savings were not recycled for farm investment through banks or credit institutions.

After Independence government reports in 1923 and 1926[32] showed that development loans were not available in Ireland. The Agricultural Credit Act, 1927, founded the Agricultural Credit Corporation to fill the gap. It was to lend 'principally to co-operatives'; secondly, to individual farmers. It did not fulfil its functions. Section 397 of the Banking Commission report of 1938 found that the ACC had loans to only eight credit coops, four of which were in default, and to eight other societies. Things had not greatly changed. John O'Donnell, manager of the Town of Monaghan Creamery, and Louis Smith went to the ACC in 1952 to discuss the society's application for a loan to build a new creamery. They were lectured by the chief executive on the unreliability of the co-operatives 'whose books were thrown in the ditch when the secretaries absconded to America'. The Town of Monaghan co-operative was established in 1901 and always had a conservative financial policy. The money was obtained from commercial sources.

Gilmore found that ACC had only 12,000 loans outstanding. Much of this was in small short-term amounts. Its method of operation was unsatisfactory. Its only office was in Dublin. It was impersonal in its dealings and its reliance on reports of gardaí, 'who no matter how well intentioned, are not versed in agricultural credit needs, farm debt paying ability, etc.' was resented. Besides, 'red tape and delays were repeatedly mentioned by farmers as reasons for not applying to the Corporation.'[33] The management of ACC distrusted the credit worthiness of farmers, giving Gilmore the example of the farmer who borrowed to buy stock but used the money for his sister's dowery and then came to borrow more for restocking.[34]

The Agricultural Credit Act, 1960, increased ACC capital by £2m. However, there was no provision for the development of direct farmer representation or of Gilmore's proposal for gradual transfer to farmer control.

In practice overdue overdrafts from commercial banks became the chief source of long term loans. In principle they could be called in at the wish of the lender, in practice they were let run.

Land as Security

Long-term mortgage, the cheapest form of credit abroad, was not on offer in Ireland. Ownership of the land, achieved in the last century, was in some ways a Pyrrhic victory – obtained at a very heavy cost. The annuities repaying the land purchase price was a first charge on income, to be met before productive investment. It was not possible for a farmer to rent land, freeing his own capital for stock and equipment.

The Banking Commission of 1938 (para. 328) pointed to insecurity of ownership as a block to credit. The Land Commission had wide powers of compulsory purchase. Furthermore payment was made in government bonds at face value, which could only be sold at a discount by the farmer.

Weakness of title was a further complication. Sons succeeded fathers of the same name for generations without bothering with the formality of proving wills and bringing title up to date. By the time the NFA brought pressure to regularise the legal situation it was estimated that up to 40 per cent of farms had faulty title.

Naturally, lenders concluded that Irish land was a poor security on which to make a loan.

Mechanisation of farming was facilitated by hire purchase. Designed to promote durable consumer goods sales to wage earners without capital assets, its high interest rate reflected high risk. The rate was concealed by quoting the interest as a percentage of the initial sum lent. Since the capital was paid by instalments the average loan over the period was a little less than half the sum first borrowed. The real burden was more than twice the quoted interest rate of 9 per cent. It compared badly with the banks' overdraft rate of 6.25 per cent.[35] The farmer took these loans because they were the only medium term (five year) loans easily and confidentially available to him.

In using hire purchase the farmer lost the discount of 10 per cent usually available to cash purchasers. The tractor salesman gained a commission from the hire-purchase company. The *Financial Times* (17 February 1964) found that these introductory commissions had run to '40 per cent and more' of the annual charge, though 10 per cent was agreed as reasonable by the motor dealers association. As the paper observed, 'it was generally recognised that these payments had little – if any – economic justification ... The general run of finance houses remain convinced that this is a thoroughly unsatisfactory state of affairs and would certainly have welcomed assistance from the government in stamping out the evil.'

Co-operative credit societies, so common on the Continent, failed in Ireland, as described in Chapter 1. A stagnant and unprofitable system of farming offered poor prospects for profitable investment till the 1950s. In December 1958 farmers were net lenders to the banks to the tune of £75 million. They became net borrowers about 1970.

Profitable farmers could obtain over-draft facilities. When farming was profitable for a short time in World War 1 the banks were over eager to lend. Credit was given for the co-operative creameries with little security. These advanced goods to milk suppliers against the milk check.

As soon as the NFA could propose a system of credit for stock purchase in the 'Heifer Scheme' money was available with embarrassing alacrity. When the Gilmore Report appeared the main banks quickly followed the recommendation that specialist officers be employed. In fact AIB made an appointment in advance of the report. When membership of the EEC opened the protected markets of

Europe, the problem (as in 1919) was to restrain an over-expansion of credit. AIB's loans to dairy co-operative rose from £5 million in 1966 to £100 million in 1976; to £240 million by 1985 and £400 million by 1995. Meetings between co-operatives, NFA and Bank Standing Committee discussed ways of limiting credit, particularly for redundant cattle marts. Creamery consolidation, on the lines proposed by Dr Knapp and elaborated by the IAOS, was financed by banks. To protect the investment AIB, main banker to all the 'Big Five', funded the weekend courses for the consolidating creameries. The Bank of Ireland endowed the Centre for Co-operative Studies in University College Cork.

The 'Green Book'

A group of NFA officers and consultants (about twelve people), worked round the clock from Friday afternoon to Monday morning in the Glenview Hotel, Wicklow, to pool their ideas in formulating an NFA policy for development. The 'Green Book', *Farm Income and Agricultural Development 1964–1970* [36] was the result. The credit section estimated capital requirements of £308 million, of which they estimated £207 could be generated within farming.

It was easy enough to see what was wrong. It was less easy to take action to right the situation even by small steps.

Heifer Credit Scheme

Robert Barton, first chairman of the ACC, had described how herd expansion was financed in Argentina, using cattle as security for the loan. Under Irish conditions this was impossible. Cattle sold in a fair were untraceable.

Eradication of TB gave each beast an identity. The NFA, in consultation with the Department of Agriculture and the Veterinary Association, put together a proposal that loans should be made on cows bearing the individual earmark of the TB test. The only security required was deposit of the Veterinary Certificate (a yellow card issued by the Department and signed by the vet) indicating that it had passed the TB test. The cow could not be sold without handing over the yellow card.

The target was to double cattle numbers by enabling farmers to hold their own heifers for breeding and farmers with surplus grass to buy breeding stock.

Sean Healy and Louis Smith set out to sell the scheme to the commercial banks. First stop was the National Bank. The chairman, John Leydon, was with difficulty restrained from immediate publication of the scheme as a special National Bank offer 'next Monday'. They accepted his invitation to lunch in Jammet's, Dublin's best restaurant. With National Bank acceptance the other banks were quickly recruited. In the Munster and Leinster there was also a heavy lunch. Full of confidence, the scheme was explained to the chairman and the board, indicating a total credit requirement from the bank of £10 million. The

Board withdrew for a few minutes and returned to say that they could do no less than the National.

This gave farmers a loan of two-thirds of the heifers' value at normal overdraft interest. Cash would be advanced before purchase and repayment would begin after two years.

In effect this was a hire purchase agreement at bank overdraft rates, with a moratorium on repayments. It gave the farmer cash to seek the best terms from the market. Had a similar arrangement been available for machinery when farmers were re-equipping themselves after the war many millions would have been saved.

Farm Plans

Merchant credit and Department grants were available for uncoordinated specific projects. Sometimes the farmer could get useful help for land reclamation and fertilisers, but not for stock to bring profit from the grass. The NFA Survey of farmer and adviser opinion bore out the judgement of Gilmore that an overall farm plan, backed by accounts and an estimate of profits, was required.

Few Irish farms kept accounts before the War. Pioneering work in pilot projects had been done by Professor Murphy, of UCC, and Dr Bob O'Connor. Helge Larsen was invited from Denmark in 1954 for a lecture tour describing the systems used in his country since 1900. From these beginnings grew interest in accounts as a management tool and as a guide in approaching banks for loans. Individual farm accounts become more valuable when compared with other results. Accounting groups were formed, for example, in the Dungarvan area. The Farm Management Surveys of the Agricutural Institute and private consultants' work for individuals developed, setting norms by which projects could be judged.

Irish Farm Accounts Co-operative was formed in 1980 by IFA, with Donie Cashman as chairman.

Bank loans doubled to £36 million between 1958 and 1963.

Gilmore's comments on the banks' lack of agricultural expertise and of contact with the world of progressive farmers were taken to heart. AIB appointed Denis MacCormack and recruited Lorcan Blake of IFA staff as agricultural credit specialists; the Bank of Ireland recruited John MacCarrick, afterwards secretary of ICOS. They also strengthened the representation of farming on their boards.

It was hard for banks to refuse loans to solvent clients. Having made farm borrowing easier, the next NFA task was to restrain over eager lenders. As early as 1958 the Banks' Standing Committee, Dr Greene, Sean Healy and members of the executive met in NFA House to warn against over-easy credit. Willie Bulfin, of Offaly, was particularly concerned at the dangers. The NFA met the banks a number of times to urge restraint in credit for superfluous marts. It was agreed to have a special line of credit for IAOS/NFA sponsored marts.

The farm balance sheet changed beyond belief. In 13 years, 1970 to 1983, assets grew from £2,550 million to £18,450 million and net worth from £2,453 million to £16,904 million.[37] The boom in land prices accounted for £12,000 million of the increase and inflation boosted all stock prices, while debts were measured in devalued pounds – just the reverse of the 1930s. The bottom line was that the farmer had become credit worthy. Bank borrowing increased fifteen fold to £1,077 million and his borrowing from the ACC twenty-fold to £469 million. Credit from Co-ops and merchants was not estimated for 1970 and 1983. Gilmore estimated it at £51 million in 1958, more than twice that of credit institutions.

Table 8 Balance Sheet of Agriculture

£ million	*1958*	*1970*	*1983*
Assets			
Land	500	1,900	15,250
Machinery etc.	40	125	1,500
Stock	220	475	2,500
Other	125	50	200
Total	885	2,550	19,450
Liabilities to banks and ACC	24.5	2,453	16,904

Source: Gilmore, and Sheehy & O'Connor, p. 39

By coincidence there was a world over-supply of savings available for investment, making borrowing seductively easy. Oil producers formed the cartel, OPEC, to treble the price of oil. In 1974 their surplus money was £80 billion, enough to buy out all the New York Stock Exchange. These funds, deposited in banks, became available for loans. However, the high price for oil caused a business recession, reducing demand for capital. With money, as with cattle, an increased supply with shrinking demand makes for low prices. With overdraft costing 7 per cent per annum and land prices rising 20 per cent it was hard for banks to alienate good customers wanting to borrow for land purchase. Land prices went through the roof when several farmers went to a farm auction, each armed with bank loan facilities.

The NFA first pointed to the need for borrowing and found the credit institutions responsive. Then it warned of the dangers of too easy a credit expansion. Much of the future work of the Credit Committee was taken up with sad cases of farmers who borrowed too freely and were caught with rising real interest rates and falling prices for produce and land.

Farm Insurance

In the shade of an orange tree in Tunis, drinking the juice pressed by the owner, were Professor Constant Boon, of Louvain University, president of the Boerenbond Belge, and Sean Healy. It was 1967. A break from the meetings of the International Federation of Agricultural Producers' conference gave time for renewing the friendship developed at many meetings and already bearing fruit in the joint venture of the FBD the year before.

Next day they were joined under the same tree by TJ Maher and Rickard Deasy to make solid proposals for the Farm Insurance Company – the pipe dream of the day before. In the New Year of 1969 £100,000 of capital had been collected, in small sums from every area. In May 1970 the Insurance Company was operating. The Irish farmers owned two-thirds of the venture. Assurances Boerenbond Belge subscribed the remaining one third and gave the backing of its £10 million capital. Its experience in administration and operation was made available by the secundment of two of their senior staff.

The development was unusually pleasant and smoothly successful. In fact it is surprising that this development was so long delayed.

Farming is a risky business, where fire, disease or accident can wipe out a family's livelihood. Macra had introduced a farm insurance agency before the NFA was formed. Tony Sherry, in his annual report of 1956 had pointed to this need, and to its potential in attracting members by the service. In July 1956 a NFA deputation spent two days in Stratford-on-Avon with the managerial staff of the NFU Mutual Insurance Company. Started in 1910, the company had a premium income of £5 million. It served half the membership, with a slogan 'Our insurance saves you twice your NFU subscription'. Its principal benefit was to the Union. Branch and county secretaries were paid £150 to £250 a year by the Union, insurance commissions of 10 per cent gave them at least twice as much – even £3,000 a year in some cases[38] Farmer controlled insurance was operating in Germany, France, Norway, Sweden as well as Britain. Juan Greene feared that this venture was beyond the financial strength of the new association.

Sean Healy again raised the necessity of farm insurance cover at the National Executive (28 January 1959) following which there were meetings with the Mutual Insurance Company, particularly concerning the risks for cattle bought on credit under the Heifer Scheme of the Livestock Committee. For this reason the Credit Committee handled insurance till its growing importance was recognised with a special Insurance Committee in 1961,[39] which arranged discounts for members through a brokerage firm.

Insurance in Ireland was expensive by continental standards and scanty in cover. The cost, about £2.7 million a year in 1961, was hardly considered in farm costs.[40] Only 70 per cent of farms had Fire Insurance and 40 per cent had third party insurance. Cover for stock or crops was rare (2 per cent). The Industrial Committee had noted the high cost of tractor insurance compared with Northern

Ireland in their submissions of 1956. Comparisons with other countries in 1961 showed Irish fire insurance approaching twice the cost in EEC countries (though other countries' buildings were wooden and Irish were of non inflammable materials). On the other hand tractor insurance was relatively cheap, with third party about one third of that on the continent.

Raising the £300,000 capital required for FBD took piority over other work. John Howley was seconded full time to insurance work in 1961. When the FBD was ready to launch in July 1969 the work intensified. Collection of NFA membership fees fell because insurance promotion took 50 per cent of organisers' time.

The value of a farm insurance service was not limited to the price. In the longer term the Association was repaid. Joint offices of union and insurance company in provincial towns were a point of regular contact for union officials with members. Office costs were shared and insurance commissions were an additional source of income.

Farm Apprenticeship

Young, trained people could not get a foot on the farming ladder. Seeing no chance of getting a farm, a considerable proportion of the few qualifying in agricultural colleges used their education as an escape from the land.

Land could only be acquired by patrimony, matrimony or parsimony. The young heir had to wait till both his parents died to claim ownership. He could not marry till ownership was secure, so the average age of farmers on marriage was 35.5 (for large farms 37.7).[41] Since the farmer's bride was on average 29 years old she usually outlived him by 10–15 years. A survey by NFA branches in Wexford showed one third of farms held

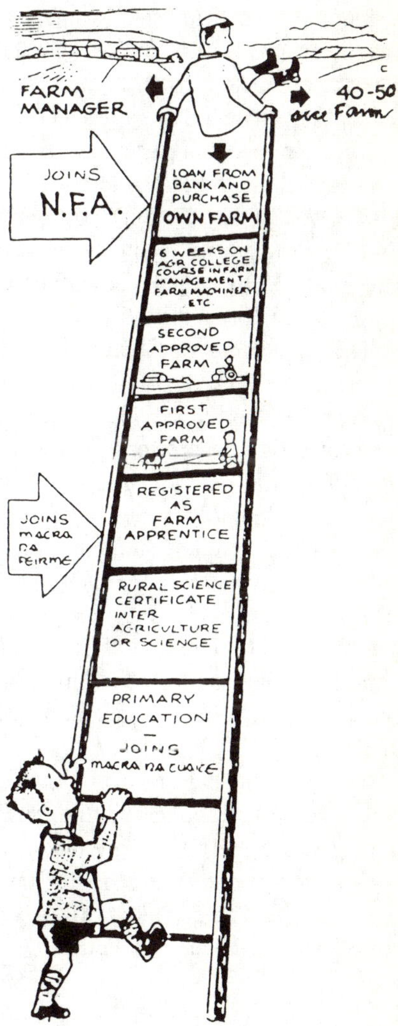

17 Farm ladder

by widows. Ownership lay with people passed the age for energetic management. Their farming knowledge was out of date.

Land-hungry small farmers bid up the price of a few extra acres, which could be cultivated with little extra cost. The price of land was above what could be paid by a new entrant to farming.

The Farm Apprenticeship Scheme put forward by the Joint NFA/Macra Committee was to combine theory, learned in agricultural colleges, with apprenticeship to leading 'master farmers'. Apprenticeship lasted four years, with at least two approved farmers. Attendance at winter courses and a six week residential course in Farm Management and Farm Accounts was prescribed. Seventeen per cent of their wages must be saved towards capital for a future enterprise. By a quirk in the regulations farm apprenticeship did not qualify as 'education' to be aided by grants or FÁS employment schemes.

It was an essential part of the proposal that the Land Commission make some farms available to the graduates of the scheme and the banks give them preference on loans. The Committee supported the proposal of Dr Philbin, Bishop of Clonfert, for a Land Bank to be reserved for young, qualified farmers as an interim measure when presenting the Farm Apprenticeship Scheme to the Minister in February 1960.

The Government's Capital Investment Advisory Committee recommended: 'the lands held by the State should be rented where suitable on short leases to young men who show evidence of competence and zeal'[42] and, 'the State should direct attention towards productive kinds of agricultural investment by providing for differential rates of interest on the loans which it makes or guarantees'.[43] Erskine Childers, Minister for Lands, changed the law to secure the rights of land lord and long-term tenants, guaranteeing that the Land Commission would not use its powers of compulsory purchase against efficient tenants. It made no practical difference to ingrained custom. Owners of land feared to make even medium term lettings and politicians feared the loss of small farmers votes if the Land Commission provided land for the landless, however qualified. Scholarships of £500 were made available and the banks were co-operative in granting loans.

Though land did not become available the apprentices earned agricultural wages while in training and, the crucial test in the market, earned more as farm managers than university graduates.

The administration of the apprenticeship was confided to Macra, aided by ICA, NFA and Department officials meeting in the Farm Centre. By 1964 300 apprentices had been enrolled.

PUBLICITY

Influence is founded on publicity. An outside observer stated: 'NFA is now one of the most effective and sophisticated pressure groups, probably in Europe ... Its research, expertise and public relations are as good, if not better than, most government departments.'[44]

Members cannot be attracted to an unknown organisation. The working farmer will not renew his subscription unless he is continually assured that something is going on. Critics of any young association question its membership to judge its importance.

The NFA gained from its parent. The 470 Macra clubs and their 7,000 members had been a sounding board through the whole country at its foundation.

18 Prosperity poster

The long term problem was to keep the individual members, scattered all over the country in their 900 branches, informed and interested. The idea that nothing is being done 'above in Dublin', or that time is wasted in foreign travel, saps confidence. Membership of commodity committees was representative of interested counties. Branch representatives at the county were to bring discussion to parish level and inform Headquarters. A large part of the 100 letters received daily conveyed branch opinion and angry comment if branches did not get immediate reply or disagreed with some Headquarters statement.

This fundamental link could easily be broken. Attendance at union meetings easily drops to the few enthusiasts. Members had to be followed home.

Local papers, read from cover to cover at the weekend, were often the only source of outside contact to the farm. The friendly editorial policy of the *Anglo-Celt*, *Leinster Leader*, the *Nationalist* and so many others was a constant strength – fostered by the *Weekly Bulletin* from Headquarters (from 1960) and by briefing sessions. George Duncan's photo-cover of NFA and Macra functions projected a positive image.

National papers had to be wooed more carefully because their readers were mostly urban. They were not hostile, but there was a wall of indifference to be penetrated. Dr Greene made good contacts with editors. That was of little avail. The practical decisions on what to print were made at the deal tables of hardworking sub-editors and reporters.

Sean FitzMaurice rapidly learned how to work the system. News rooms are flooded with press releases from many sources, gaining minimal coverage. Attention is paid to the aimed shot, not to the shot-gun blast. The time of delivery had to be right, targeting country editions or the Sundays. Presentation had to be in double spacing, with wide margins and subheadings to facilitate the reporters' work. Reports were varied for each paper, lest competing papers carry exactly the same story. Above all, trust in the reliability of information had to be built and the assurance that a phone call to the NFA at any hour would get a quick, informed response. A readiness to go along to the editorial office for discussion and to check proofs made work easier for both sides. Calls from papers in London and New York had to be serviced at any hour.

A couple of hundred cigarettes to the reporters' room at Christmas (an unexpected gesture, too little to be a bribe) was no harm.

The influence of radio and, later, television could not be ignored. Mr Rughe-imer, a Swede, was the first Director of RTÉ. Accustomed to farm organisation, he was in contact with Sean Healy as NFA general secretary twice a week. Paddy O'Keeffe and Michael Dillon, among others, could be relied on to give coverage to the Association's view and opportunity to its spokesmen. The news room presented difficulties. The staff were newsmen like those in the daily papers. They recognised a good story and knew the interest of the public. They had sympathy with the underdog seeking to state his case. However, being in a government monopoly, they were not free agents with final say on topics broadcast. Government objections that NFA releases got priority coverage over the Minister's led to cut-backs. This was partly evaded by arrangement with the news room. Controversial NFA releases were delivered so close to news time that consultation with the Government was not possible. The NFA report would receive coverage for one reading on its merits as news. In the next newscast the Government version would be given, at times without the original NFA story. During the Farmers' Rights march there was a marked difference between the coverage of RTÉ and BBC. The Donegal marchers designed their schedule to spend extra time in Northern Ireland to avail of the more favourable BBC coverage. Government censorship had been even stronger in the milk strike of 1939. Then the first use of the Emergency Powers Act was to block, by censorship, all newspaper reporting of the farm dispute. The LMP secretary had to travel North with copy for the Belfast papers to print the story.

It is impossible to overestimate the impact of the *Irish Farmers' Journal*, carrying independent technical and market knowledge together with the Association's doings. The *Journal*'s influence was based on its independent comment on all affairs. Both *Journal* and the NFA would have suffered had it became a mere 'house magazine' of the NFA.

Independent comment in the Journal was essential for credibility but sometimes unwelcome. Editorial comment did not always suit the NFA executive. From 1961 the *Journal* printed headed pages supplied by headquarters of the NFA and Macra each week. Even then freedom of the press could lead to friction. Feeling ran so high during the Farmers' Rights march that copies of the *Journal* were burned in Tullamore by angry farmers. The Department, equally incensed, forbade all advertising in the *Journal.* A newsman's lot is not a happy one.

The *Weekly Bulletin* took three or four topics each issue. It was particularly valuable in briefing branches and local papers during the dispute of the 1960s, when it had reached its 600th issue.

The *European Bulletin* was issued by the Brussels office from 1973.

The *Yearbook* gave an annual report of each committee and covered some burning topic such as Marketing (1961), Agribusiness (1964) or Small Farms (1965). The NFA Note-book was to be in the members' pocket and the *Yearbook* on his mantle-piece beside *Old Moore's Almanack* and the creamery book. Production was irregular (or copies have not survived) till 1969.

Conferences or conventions were to build morale and gain general publicity every five years or so. The Farmers' Pavilion at each RDS Spring Show was a place of reunion.

Publicity costs little. It only takes skill to know a good story and to be first to the press with a comment. The NFA coverage was second only to the Government. Little was paid for advertising, the material as news worthy in its own right. The section was almost self-financing with the profit of the *Year Book*.

7

A Foreign Policy for Farmers

The Anglo-Irish Trade Agreements failed for agriculture. This note constantly recurs in the NFA records. The dead meat trade, giving employment in meat factories and by-product processing, was stifled by the advantage given to store cattle. Even the store cattle trade offered no secure outlet, as was shown in the glut of 1966.

Juan Greene wrote: 'Our advantages of Imperial Preference are more than compensated for by these countries' lesser transport and other costs. Comparing the position of a particular product exported both from here and Holland with a Dutch friend I discovered that our so-called 10 per cent advantage of Imperial Preference (now nullified in pigs, canned cream and blue cheese) turned out to be almost a 20 per cent disadvantage when considering all factors.'[1]

The message was driven home in November 1961 when Britain imposed quotas to regulate her butter market. Irish foreign policy had failed to win access to markets, while internal policy had raised costs of production and ignored the requirements of marketing (see chapter 5).

THE CONTINENTAL MARKET

The first NFA demand for a more active foreign policy was dated 4 April 1955. The Government immediately saw an important role for NFA in trade development. In September 1955 Sean Healy was asked to represent the Association at a conference in Cologne, for which the Deparment would bear 75 per cent of the cost. A few months later, in April 1956, the NFA National Council discussed a memorandum: 'What We Would Gain from a United Europe'. The partnership of farmers and Government was explicit at the meeting with Sean Lemass in 1957.

In 1957 the text of the proposed Treaty of Rome were translated for the Council, together with an analysis of the alternative proposals on offer in EFTA or in isolation. The conclusion was that the Free Trade Area with Britain gave Ireland the worst of both worlds. Ireland should make membership of the European Community its objective.

Excitement stirred at the prospect of a common agricultural policy under the

Treaty of Rome (1958). The Treaty proposed the same prices to all member states, with a Court to adjudicate disputes and a policy of raising farm incomes to parity with the rest of the community. NFA memoranda multiplied on the topic.

The EEC proposals for agricultural policy were published in Brussels on 30 June 1960. On 20 July the NFA National Council resolved :

> The Council of the NFA, having before it the Report on 'Agricultural Proposals Submitted by the Commission of the European Economic Community (EEC)' laid before the Council of Ministers on the 30th June 1960:
>
> Believing that substantial progress is being, and will continue to be made during the next few years in establishing a common agricultural policy for the EEC covering a wide range of agricultural products, many of which are of special interest to Irish producers:
>
> As a consequence of our failure to obtain a fairer deal for our agriculture in the recent Review of the Anglo-Irish Trade Agreement:
>
> Conscious that we should now make a serious and thorough examination of policy in consideration of becoming a member in the EEC:
>
> RESOLVED that the Irish Government be urged to seek immediate clarification and inform the public on these vital points which are now all the more urgent and pressing because of Britain's and EFTA's negative policy towards our very serious and depreciating agricultural position.[2]

The Economic Section was asked to prepare a full report on the agricultural proposals. In seven weeks, with a visit to Brussels, a booklet[3] was in the hands of the members. It set out the farmers' case for membership, the proposals for each commodity and for the Social and Structural policies . Two printings sold out in weeks. Introducing it Dr Greene wrote: 'In the event of the UK not joining the EEC, the farming interest would recommend our Government to reserve the right of independent decision, following on a new assessment of the situation. We would certainly consider it a mistake to say at this moment, that if Britain does not join, we will not join either ... It would be foolish to regard the European Common Market as a farmer's 'Utopia' but it is as close to it as our present position is to a farmer's 'nightmare'. Many quotations could be produced giving this message. Dr Greene made headlines, lecturing at Inch Folk Group. *There is no utopia for agriculture in the Common Market* ... At all costs we must avoid the Common Market being used as an excuse for inaction or becoming a fashion as the cure-all for our agricultural ills; there is no Utopia for agriculture ...The Common Market will provide us with a greater opportunity. It is for us to grasp and exploit to the best of our ability the opportunity it offers'.[4] NFA was not starry-eyed, but saw the main chance in the Common Market. Impatience with Government's inaction would have exploded whether Greene, Deasy or Maher were president.

In 1961 the ICMSA commissioned the economist Dr Raymond Crotty to brief their National Council on the prospects for Irish agriculture in the EEC.[5] He foresaw a drop in creamery milk price during the transition period and an increase of about 30 per cent on attaining full membership. 'The price should reach 12 pence a gallon compared with 9 pence in 1961.' He warned against stimulating milk output as it might jeopardise chances of Ireland's acceptance as a member. Beef should show gains similar to milk. 'Present indications are that the continental tastes are veering towards an appreciation of fatter, better conditioned cattle and away from the leaner type, which are essentially a by-product of dairying.' But: 'However remote it may appear at the moment it is quite probable that of all sectors of Irish agriculture egg production will show the greatest expansion in response to membership of EEC.' However, the general conclusion was that: 'It is inconceivable that any net losses would offset the gains forecast for the principal items, beef, manufacturing milk and sheep. Clearly, on the whole, Irish farmers stand to benefit from Irish membership.' Because Irish industry would be uncompetitive, requiring increased taxation of farming for support. 'The picture of the overall economy is of a declining industrial arm combined with a more prosperous, but undynamic agriculture. Ireland's freedom of action is confined to the rejection of the less attractive option, that of remaining outside EEC.'

Dr Crotty became a main contender against membership in the referendum debates.

In 1972 the ICMSA reached a unanimous decision to support entry to EEC.

For the immediate future Britain was the main market open. The Common Market was only a proposal which, like several others in the recent past, might come to nothing. Asher Winegarten, chief economist of the National Farmers' Union of England, in conversation with Louis Smith, said, 'Britain should have joined the Community at the beginning, when it was only an agreement in principle. Then it would never have happened.' For Ireland to have joined the EEC would have risked damaging relations with Britain before compensating markets opened on the Continent and while Irish industry was still uncompetitive. It might have been possible. M. Pisani, the French Minister for Agriculture, said that his veto on Britain's application was essential in 1961 because they would have blocked all agricultural market developments. Any other member would have felt compelled to impose the veto if France had not done so. Both Denmark and Ireland would be useful allies to France. However, Denmark could not be accepted without Britain because her production was large enough to upset the EEC market. Ireland would be a useful ally of France and so small in output as to have insignificant effect on the market (conversation with Louis Smith).

European Free Trade Area

The European Free Trade Area was dismissed by the Irish as offering no benefit to agriculture. This was a little hasty. As the price of entry Denmark obtained : from Britain equality with Ireland on products such as cheese and, in addition, the right to appeal against any new internal measures injuring their trade; from Sweden a refund of 60 per cent of tariffs levied on food; from Switzerland an undertaking that 40 per cent of agricultural imports would come from Denmark; Austria, Portugal, Norway gave Denmark preferences. The Danes negotiated agreements with Russia and Germany by which they undertook to make imports from Denmark balance exports.

An NFA article on our trade policy in the light of the Anglo-Danish agreement was headed 'Now We're F...ed'. 'While we were fighting the last? battle of the Civil War, the Danes have come to England and gone home with the bacon, the cream and apparently would have got the jam if they had asked for it. After the door had been opened to let Danish pigs into our protected market we sent, not a full deputation to negotiate, but merely a couple of gentlemen for preliminary talks.'

Diplomacy

The NFA repeatedly expressed concern at Irish diplomatic policy. Till Sean MacBride's advent as Minister for External Affairs the objectives of Irish foreign policy were almost exclusively political. Embassies were active in culture but little involved in trade. MacBride gave a new development to Irish diplomacy. He was active in the preliminary negotiations for the formation of the EEC. With his French education he fitted easily into the society of the continental politicians. He even appointed an economic adviser (Louis Smith) working on trade liberalisation for a short time in 1949. Under Frank Aiken the focus of diplomacy shifted to the world stage. A high profile in the United Nations gives an ego-trip to a country of 3.5 million people. It may be helpful to the world but does nothing for welfare at home. Only in London was there an Agricultural Attaché. Brussels was treated as a minor embassy, its two staff could not possibly cover the development of the EEC. An Agricultural Attache was only appointed in 1962 – after the first application for membership had failed. It was quite impossible for him to cover all meetings and arrange programmes for visiting delegations. The Danes, with contrasting priorities, appointed an extra six attaches as part of the settlement of their commodity strike of 1961.

Examples of flacid trade policy were not hard to find. The Department of Agriculture told an early NFA delegation that they had not attended a conferences on trade, 'because the British kept us informed'. The British, with the best will in the world, saw things from an opposite point of view to the Irish.

When the French closed their lamb market in 1962 the Dutch reacted within hours to regain entry by retaliation against imports from France. The French farmers association explained to the NFA that there was no special animosity against Ireland, apart from bad grading. Ireland, 'had not negotiated'. Diplomatic protests without either threats or promises were useless. As France's second best customers per person for brandy, with whom France had an overall trade surplus, Ireland had the power to retaliate. The NFA did ask its members to refrain from buying French cars. It would have given a false impression of farming prosperity to suggest a boycott of brandy.

Some Irish lamb was still able to enter the French market, but on the Dutch quota.

A German trade mission called to NFA House, frustrated that they must go home without giving an opening for Irish exports. The Danes made a trade agreement that Germany would import more food to balance its trade. A similar agreement was open to Ireland. However, since the 1964 Anglo-Irish Trade Agreement specified that Britain should have a more favourable tariff position than any other country, there was no deal possible for Germany. Any concession they obtained would give greater advantage to Britain.

As each trade question arose the NFA urged that a commercial line be considered. In aid to the Third World Ireland contributed cash. Other countries tied their aid to purchase of their own products. NFA wrote to the Red Cross that Irish aid should take the form of surplus food. The Red Cross replied that purchasing decisions were made in Geneva, not in Ireland.

When Irish troops went to the Congo no provision was made that their food be Irish. As far as Ireland was concerned a market of 1,500 men 'in the prime of their eating' valued at £300,000 was lost. Only their emergency rations, sought after by other contingents because of their quality, were of Irish origin.

Though Ireland had a deficit with all European countries except Britain this trade leverage was not used. Because Ireland did not recognise the Soviet Union no trade agreement was possible. The Soviets balanced our imports with British exports.

The NFA constantly drew the Government's attention to its diplomatic inaction. In November 1961 the reports of other countries' activity, gained in the IFAP, were sent to the Taoiseach, also the reports of the Danish Agricultural Attachés. The Irish Department of External Affairs increased their expenditure on culture by 30 per cent in 1960 but could find no money for an agricultural attaché. Each NFA committee drew attention to the need for export quality standards and for marketing bodies similar to those in competing countries.

The European Movement

Forty odd years on few remember the European Youth Campaign. It aimed at wakening among those in their early twenties a feeling for Europe. It followed the Hague meeting, at which the European Movement was founded. This age group was very sensitive to the slaughter of their generation in the World Wars, which were civil wars of Europe. The poverty resulting from the economic disruption of war and the protectionism of the 1930s and 1940s was only too apparent.

Older Irish people, particularly the politicians, were so entangled in a love-hate relationship to Britain that their attention could not be captured by developments outside the English speaking world. The Council of Europe listed 85 international, nongovernmental organisations promoting European union in 1954. Only one had an entry of Irish membership.[6]

In Macra admiration for the success of Danish farmers in education, co-operation and farming kindled interest. By beating us in the British market they forced themselves on our attention. It was through the European Youth Campaign, with funds from the Council of Europe, that lecturers such as Helge Rassmussen, Larsen, Per Federspiel (chairman, Agricultural Committee, Council of Europe) and others could be made available to Macra and the IAOS.

The NFA's pro European Common Market policy had to be sold to the Irish Government and public. The Irish Council of the European Movement provided a platform. Meetings, subsidised by Esso (Ireland), were held throughout the country. Prominent figures were invited to speak in Ireland. At a meeting at the Shelborne Hotel, Asher Winegarten, chief economist to NFU, explained why the British farmers would prefer not to join EEC; Willie Biesheuval, president of the Netherlands' Farmer's Union and later prime minister, offered the Irish 'a window on the Common Market' – to be told, with loud applause, that it was not a window that was required, but a barn door so that we could drive cattle in'.[7]

The European Youth held their meetings in the first office of NFA, No. 24 Earlsfort Terrace, sharing the kind hospitality of the RGDTA boardroom. When No. 27 was taken over, Denis Corboy occupied the basement as Secretary of the new Irish Council of the European Movement. In Belfast similar facilities were given by UFU. Louis Smith succeeded Garret FitzGerald as chairman. It was very hard for the European Movement to attract the attention of politicians to European developments; businessmen were receptive.

Referendum Campaign. 1972

In preparation for the Referendum on accession the National Council appointed Joe Rea, later president of the IFA, chairman of the National Referendum Committee. There were IFA directors of election in every county, with instructions: 'IFA is to be totally, completely and entirely committed to a very active campaign for Irish membership of EEC and for the greatest majority possible in favour of "yes" in the referendum.'[8]

In preparation for the Referendum on EEC membership a short booklet was prepared by Bindon Blood, Alan Dukes, Paddy O'Keeffe, Louis Smith, and Michael Sweetman, to bring up-to-date information on the Irish prospects for the food industry.[9] It was easy to convince farmers by economic arguments. There was no way that the home market could absorb present production, never mind the increased output which modern methods would yield. Food processing could not develop in Ireland while Britain had lower prices of raw materials and the Europeans charged progressively higher tariffs at each stage of processing. It was less easy to convince the consumer that beef at European prices was in the national interest. If the beef price rose 70 per cent, but 80 per cent of output was sold abroad, more equipment for factories and more consumer goods would be available in Ireland. Every urban constituency gave a majority assent.

INTERNATIONAL FARM ORGANISATIONS

Sean Lemass had stressed the interdependence of the Government's diplomatic activity and the participation of voluntary associations in developing Ireland's international status and trade policy (see chapter 5). A proper area for NFA activity was in the international associations of farmers formed after the War. The Federated Farmers of Ireland had been formed with these contacts in mind. In 1950 the IAOS had been approached to represent Ireland but its interests were at that time too narrow to warrant the expense. Almost all our agricultural exports went to Britain, making European and world considerations of minor interest.

The International Catholic Rural Association meeting in Rome, called to assess the implications of the recent papal encyclical *Mater et Magistra*, was attended by Rickard Deasy, Tom Cahill and Sean Healy, as described in chapter 1 (see also page 75). The ICMSA continued its interest in ICRA, of which Jerome O'Shea, general secretary of ICMSA, became general secretary.

In 1956, just a year after formation, the NFA examined the international organisations. Though invited to join the Confederation of European Agriculture (CEA), NFA decided that the members' interests would be best served by participating as a full and active member in the International Federation of Agricultural Producers. Formed in 1947, with headquarters in Paris and 130 million members world-wide, this organisation covered all matters of farming interest, production, marketing, finance and trade. It had consultative status with the OEEC, EEC, United Nations and its subsidiary the Food and Agricultural Organisation. IFAP was preferred to the CEA, which represented the more protectionist views of European landowners. As Roger Savary, secretary general of IFAP, said: 'The NFA application for membership was received with particular favour because its constitution and its first achievements conformed so closely to the ideal standards ... The entire outlook of the NFA as it appears in the 1958 *Yearbook* – the understanding of agriculture's place in the national economy, the need for a comprehensive policy embracing various commodity groups, and, above all, the desire to seek solutions to trade problems through international consultation and co-operation, fitted so admirably the pattern of IFAP's views that the Irish farmers' organisation was a member of IFAP family even before it knew it.' Perhaps this was not surprising when NFA had picked the brains and copied the ideas of all the neighbours.

The NFA immediately became observers at all meetings and were accepted as full members in 1958. They made their presence felt. Juan Greene was elected to the IFAP Executive at Rome in 1962 and made vice-president at New Delhi, 1964; he was asked to take the chair for the meetings of the Policy Committee – the principal group – at the IFAP World Congresses in New Delhi, Dubrovnick and Bray. Sean Healy chaired the Organisation and Finance Committee at the Tokyo Conference and regularly in Brussels; Louis Smith chaired the Select Committee on Farm Structures appointed in Dubrovnik to draft IFAP policy on farm size in Europe (1962) and of the Committee on Produce Prices and Farm Input Costs. The NFA represented the small butter exporters at the International Dairy Congress. Information was exchanged on markets and policies. NFA committeemen had the developing experience of discussing in Paris one day and at a local meeting in Crosserlough the next.

Juan Greene's most widely noticed project was the use of the growing agricultural 'surpluses' to alleviate world hunger. A small tax on the wealthy countries could solve problems of both rich and poor. It was not to be a final solution but a way of using surpluses to fill a crying need. His proposal was first presented to IFAP in New Delhi; then taken up by the press in January 1960; it became part of his presidential address at the AGM. He was invited to explain his ideas to the Organisation for European Economic Co-operation, Paris, in May; it was sponsored by Adlai Stephenson, US ambassador to the United Nations, in March 1961.

While Ireland could not yet be a member of COPA (the federation of farm organisations of EEC), regular reports were given to the European Committee of IFAP by M. Rabot, head of the Agricultural Directorate of the Commission in Brussels. NFA was acknowledged to be better informed than the Government on the developing Common Agricultural Policy.

Under Sean Lemass the Government agreed to help finance these activities because it was in Ireland's interest to be represented. This was clearly stated at his meeting of July 1957 (see chapter 5) and put in writing more than once but most clearly in his memorandum of 1964. 'The Government will continue to give financial assistance to the NFA in sending representatives to international conferences of farmers' organisations.'[10] This promise was not honoured. The Government gave a dinner and reception to the delegates to the IFAP World Conference in Bray (as they would to any large group of tourists). When Sean Healy remarked to a Dutch friend, Rijnse Zylstra, on the heavy burden of travel costs for an island country the Dutchman was amazed at the problem. Some years previously FAO had made an agreement with governments that, if the national governments paid half the expenses of 'non governmental bodies', the FAO would pay the rest. NFA knew nothing of this. When the matter was drawn to the Department's attention the NFA was paid 50 per cent of expenses for the future. It was presumed that this was the FAO contribution, but no explanation was given. Other countries' delegates had their expenses paid in full. NFA settled for 'half a loaf as better than no bread'.

A later promise of aid for international representation was made after the Farmers' Rights Campaign. The NFA did not seek this aid. They felt bound by the resolution of the mass meeting in the National Stadium, Dublin, at which FBD was formed in 1967, that no Government subsidy should be asked or accepted.

The Annual IFAP seminar on Agricultural Co-operation was held in Dublin in 1959. The World Council met in Bray, county Wicklow in August 1960 and the European Committee in Dublin in March 1961. It was a chance to show that Ireland was a modernising country.

That this was not idle tourism became apparent when Jerome Buttimer and Sean Healy attended the European Dairy Conference at Cremona, in 1958. Ireland was not recognised as a dairy producer. It was important that Ireland was in a position to state its case. If you are not seen and heard you are ignored.

Juan Greene described the Irish Government attitude to Agriculture as 'a patchwork of partial remedies'[11] In the Trade Agreement we left negotiation too late, 'we could only search among the crumbs for anything that had been left over from the earlier feasts with the Danes and the the Outer Seven countries'.

The NFA objective in IFAP was to contact foreign farmers. They had most in common with the Scandanavians, Danes and Dutch in language and in commercial interests; Louis Smith with thhe French; Juan Greene was immediately accepted by the British and Commonwealth countries, partly because of trade contact and experience, perhaps because of his service in the RAF.

It quickly became evident that Irish interests were different. The British farmers considered their home market their sole preserve, where imports were privileged to fill a shortfall in home production. Their large farmers had little in common with the small producers of Ireland, the Continent, or, indeed, of Scotland and Northern Ireland. Members of the Northern Irish marketing boards told how they were unable to get service from the British embassy which was freely given by that of the Republic. When Harry West and Humpfrey Jameson, of the UFU, met with Neil Blaney, Rickard Deasy and Sean Healy in the Russell Hotel, Dublin, the common interest of farmers North and South, as opposed to those of England, was the subject for discussion. So it happened, one morning in Paris, that the NFU took their seats on one side of the table and found that the representatives of the Scots and Ulster farmers were seated opposite, with the French and Irish. The Dutch, Belgians and French saw Ireland as a valued ally in EEC negotiations.

In Dublin a thousand farmers attended the one day conference on EEC in February 1970, described by the *Irish Independent* editorial as 'one of the best bird's eye views of the subject for a long time.' Dr Jacques van Lierde, of the Commission in Brussels, outlined the Structural Proposals being developed. Farmers could not ignore these developments and good marketing if they hoped for farm incomes comparable with those in other sectors. Rinnse Zylstra, President of the Farmers' Union of the Netherlands said: 'Agricultural policy should not in a responsible world be a solution for one group of farmers, and the graveyard for other groups. This brings me to the conclusion that much closer co-operation among nations and among farmers and their organisations is necessary to meet these problems. That is what I mean by saying there is no national solution for international problems.'

It has been said that farmers were promised a bonanza, without warning of the need for reform. Pages could be filled with quotations from Juan Greene, from the NFA *Yearbook* 1969, from Louis Smith's speech at the RDS meeting, to show the contrary.

Comité des Organisations des Producteurs Agricoles (COPA)

The NFA had been so active within IFAP, had formed so many personal friendships, that T J Maher, as president, and Sean Healy, as general secretary, were invited to COPA meetings, with observer status and the right to speak, in 1972 – even before Ireland became a member of the European Community. No other applicant country received this privilege. The Rules of COPA in Irish were made available from the University of Louvain.

Brussels Office

On accession to EEC, Government and farmers were almost at one. Conflict on prices moved from Dublin to Brussels, where the antagonists of the 1966 Farmers' Rights demonstrations found common ground. Discussions on policy took place regularly.

Co-operation on international affairs became very close, particularly during Dr Garret Fitzgerald's term of office. It was never as close as in the Netherlands, Denmark or Belgium, where trade associations of agriculture or industry were invited to take part as members of trade delegations. Even Irish state companies, such as Bord Iascaigh Mhara, were not intimately involved. The immediate backup of people with practical experience of production and trade gave these small states the edge in negotiations with Britain. Regardless of disputes or statements outside, it was recognised that the Department, NFA and the Brussels offices of government and farmers were on the same team.

In 1966, through IFAP the first warnings of the impending treat of GATT on European agriculture were sounded in Earlsfort Terrace.

8

The Philosophy of the NFA

A durable institution needs a unifying purpose beyond its immediate, day to day activities. Members must be held together by a common aim important enough to over-ride their occasionally conflicting personal interests. Negative aspects were easily agreed: the NFA should be non-sectarian; avoid party politics; each farm must have an equal vote. Such agreement removes obstacles but does not inspire action.

At first the NFA objectives evolved by consensus on individual problems rather than by formal discussion. A unifying text came fortunately to hand in the papal encyclical *Mater et Magistra* (1961), dealing at length with social justice and rural life. Few of the Catholic farmers, probably none of the Protestant, had ever read a papal encyclical. This document, and its author, struck a responsive note, offering a philosophy applicable to Irish farming. As Pope John said to visitors from an international farm organisation, 'if I was not in this post I would be among you.' To spread knowledge of the encyclical the Archbishop of Dublin, John Charles McQuaid, contributed towards the expenses of Rickard Deasy, Tom Cahill and Sean Healy in attending the International Meeting of Christians on Rural Life in Rome, September 1962. The trip yielded a store of ideas. The group used the opportunity to make foreign contacts, visiting five Italian, two French, one Belgian, two British farm organisations, FAO, IFAP, OECD, and officials in Italy, France, Belgium, Britain. They visited Burgo a' Mezzano, birthplace of Pope John XXIII, where Esso had a rural development project. This part of the trip was paid for by Esso Oil.

The *Irish Farmers' Journal* published articles by the Revd Dr Dan Duffy (member of NFA National Council for Monaghan) in 1963 quoting and explaining the encyclical. The NFA reproduced them as a pamphlet,[1] with introduction by Rickard Deasy. Though the encyclical was written for Catholics, its message was acceptable to those of any religion or of none. Juan Greene complained that he had been saying the same things for years without response. The pamphlet was used at meetings. An early photo of the National Council at work shows the pamphlet on *Mater et Magistra* among all the commodity committee reports scattered on the boardroom table. Farmers carried pencil marked copies for use at local meetings.

In the early 1950s the IAOS had to send different organisers to serve Catholic

19 An NFA council meeting in progress

and Protestant farmers in East Donegal. They could not work together in a co-operative. The common social philosophy in NFA and co-operation with the UFU produced harmony. Sean Healy was in Letterkenny when Pope John XXIII died. He met a local Protestant farmer, who asked him 'Any news, Sean?' 'John XXIII is dead', said Sean. 'Ah. That's a pity,' said the farmer, 'I suppose they will replace him with a Papist.'

Passages of the encyclical were tailormade for NFA:

> It is important to bear in mind that in agriculture, as in other sectors of production, association is a vital need to-day- especially in the case of family farms. Rural workers should feel a sense of solidarity with one another, and should unite to form co-operatives and professional organisations. These are very necessary if people working on the farm are to benefit from scientific and technical methods of production and protect the prices of their products. They are necessary too if they are to attain an equal footing with other professional classes who, in most cases, are likewise effectively organised. They must organise, finally, in order to have their proper say in political circles and in public administration. The lone voice is not likely to command much of a hearing in times such as ours.
>
> We therefore desire to express our satisfaction with those sons of ours the world over who are actively engaged in cooperatives, in professional groups and in worker movements intent on raising the economic and social standards of the agricultural community.

> Given the special nature of agricultural produce, modern economists must devise a suitable means of price protection. Ideally, such a price protection should be enforced by the interested parties themselves, though supervision by the public authorities cannot altogether be dispensed with.
>
> On this subject it must not be forgotten that the price of agricultural produce usually represents the reward for labour rather than a return on capital.
>
> The encyclical said that 'the time has now come to promote in agricultural regions the establishment of those industries and services which are concerned with the preservation, processing and transportation of farm products. Enterprises relating to other sectors of the economy might also be established there. In this case the rural population would have another means of income at their disposal ...

Particularly attractive to farmers was the argument of the encyclical for income parity, that the interests of the consumers in cheap food, 'cannot be used as an argument for keeping a section of the population – farming people – in a permanent state of economic and social inferiority, depriving them of the wherewithal for a decent standard of living. This would be directly opposed to the common good.'

This, and more in the same vein, repeating the arguments of Sir Horace Plunkett, Fr Tom Finlay and so many others, summarised the aims of the NFA. It justified the effort to gain the right to consultation with the Government; to farmer controlled marketing boards; to price stability; to incomes comparable with non agricultural workers. They believed that the Government shared these principles. De Valera stated that Fianna Fáil policy was firmly based on the Social Encyclicals, Lemass agreed in practice. The jealous centralisation of research, education and advice, breeding, grading, marketing boards in the central government came as a shock.

General Michael Collins expressed the NFA ideal most succinctly when journalists pressed him on the object of his struggle for national independence: 'I am fighting to make the fucking country fit for adults to live in.'[2]

Religion and Farm Associations

In a country where 85 per cent of the people go to church every Sunday – in country districts a much higher proportion – relations with the churches and the social philosophy of the religious groups has practical importance.

There is no conflict of philosophy on professional organisations between the main streams of religion in Ireland. Due to the shortage of educated leaders who could be recognised as being firmly on the side of the people, clergy of one denomination or another were prominent in rural organisations. As education and experience widened, the laity took over without dispute.

The Ulster Farmers' Union had a tacit arrangement that every fourth president would be a Catholic. It seemed remarkable to no one when at several meetings Protestants represented the farmers of the Republic and a Catholic the UFU. No understanding of office sharing on religious grounds was considered in the South. Even a tacit recognition would be divisive where the members are quite unaware of the religious affiliation of candidates, where religious services are held in common. With very occasional tensions the non-denominational line was held by all the farm organisations.

As practical men the farm leaders saw the strategic value of being seen to be on the side of the angels.

Archbishop McQuaid conveyed his blessing to the Inaugural Meeting. It was common practice to ask a blessing on new ventures. the Farmers' Rights Campaign developed in 1966 A letter from Mgr. D'Ascenzi, from the Vatican on behalf of the Pope, endorsed the Declaration of Farmers' Rights issued by the NFA during the protest march,1966. It was headed by reference to the papacy and ended with the Proclamation of the Republic 1916, guaranteeing 'equal rights and equal opportunities to all citizens'.

Charity

Charity is not an objective of farm organisation but is a natural reaction of any Irish group. In 1960, for example, £30,000 was collected by NFA for the World Refugee Fund.[3] At the same time the Association itself struggled for five years to raise £10,000 to buy NFA House and, as late as 1967, the expenses of National Council Members were paid by an annual draw whenever funds became available.[4]

The Revd Livingston, head of the Methodist agricultural college, resigned from the National Council to go as a missionary to Uttar Pradesh, India, in 1965. The Council decided to support his project of livestock improvement. It became an ecumenical occasion. A collection was made by the Council through NFA branches (presumably 90 per cent Catholic) and by the UFU. Kerry and Jersey bulls were presented by the Salesian Order's Agricultural College, Warrenstown, and Jerseys by the Bewleys, who were Quakers. Livingston, impressed by his reading of *Mater et Magistra*, requested Sean Healy to arrange a private audience with Pope John on his way to India. Used in artificial insemination the five bulls served 1,000 cows a year. It was reported that the progeny gave three to four times the yield of their dames. Mr Young of Athlone, son of the NFA Council member, volunteered to travel with the stock and supervise their acclimatisation.

A similar project was undertaken sending pigs as breeding stock to South Korea under Bill Harty.

From the 1980s IFA has been active in the Grow Fund for Third World rural development, placing a voluntary levy on the grain and sugar crops. Leinster Milk Producers gave dried milk to Europe after World War 2.

Nearer home aid was given between areas in Ireland in times of crises. In

DECLARATION OF FARMERS' RIGHTS

We the Farmers of Ireland,

CONSCIOUS of the acceptance of the Government of Ireland of the social objective enunciated by Pope John XXIII of preserving the family farm;

AWARE that this aim cannot be achieved until the family is put in a position to earn an income high enough to support standards of social and educational progress, social security, home amenities and leisure equal to the rising standards achieved by comparable workers in the other sectors;

PROFOUNDLY REGRETTING that in relation to almost all of our rural population existing standards fall far short of those already achieved by the other sectors;

KNOWING that improvement of the relative social and income status of the underprivileged agricultural sector depends, not only on the operation of effective and adequate farm policies, and on the modernisation and increased productivity of individual farms – but to an ever-greater extent on the full development and efficient operation of the supply, processing, and marketing arms of Irish agriculture;

ACKNOWLEDGING that agricultural development must remain the backbone around which a more broadly based rural economic programme must develop;

REALISING that the demands of a balanced economic growth and the increasingly competitive nature of the agricultural export trade demand an unprecedented effort to develop and modernise our greatest and least developed national resource – agricultural chain;

RECALLING the as yet unfulfilled guarantees of the Proclamation of 1916 and in particular that the Republic guaranteed:–

> 'equal rights and equal opportunities to all its citizens' – and declared – 'its resolve to pursue the happiness and prosperity of the whole nation and of all its parts, cherishing all the children of the nation equally;'

CALL upon the Government to honour the promises of 1916 and proclaim and declare the following to be the basic rights of the rural and farm families of Ireland:

1 THE RIGHT to an incomes policy which effectively closes the gap between the incomes of farmers and comparable workers in other sectors;

2 THE RIGHT of small farmers to a well planned system of production incentives equally available to all potentially viable small farms so as rapidly to build up their productivity;

3 THE RIGHT of communities in underdeveloped areas of rural Ireland to a comprehensive rural economic development programme;

4 THE RIGHT to social security and retraining opportunities equally with other citizens;

5 THE RIGHT to immediate commodity price and credit policies so as to ensure the real availability, proper employment and fair remuneration of all the factors of production including working capital;

6 THE RIGHT to professional agricultural education;

7 THE RIGHT to a financial and legal framework – affording to all farmers encouragement and opportunity for involvement in the full development of the co-operative movement;

8 THE RIGHT through their representative organisations to full consultation and involvement in the formulation and putting into effect of agricultural policy;

9 THE RIGHT to protection against the effects of unchecked vertical integration or foreign control whether in the supply, processing, or marketing arms of agriculture;

10 THE RIGHT to effective participation in the marketing of agricultural produce so that from now on our output will be sold to the best advantage;

11 THE RIGHT to carry no more than our fair share of taxation and in particular the right to relief from systems of taxation such as local rates in which the burdens are not proportioned to the capacity of the people contributing.

1977 the heavy floods in the Shannon basin caused such loss of fodder that cattle starved. The IFA arranged the transport of gifts of hay from members in more favoured areas.

The Family Farm

Is farming a business or a way of life? Is farm policy a question of economics or of sociology? There is no issue more divisive of farm organisations.

The NFA, like other rural organisations, based itself on the viable family farm as a social unit giving employment to the farmer, his wife and family. Such an ideal modern 'family' farm was far too large to be a practical immediate objective in Ireland.

The NFA was probably wise to dodge the emotive issue. It had ample uncontroversial work to be getting on with. Farm incomes could be increased by raising output; by increasing efficiency in production and marketing; raising prices. Small farms stood to gain most from co-operative organisation. These measures could ease the pain but could not cure the problem of the smallest farmers. The West Donegal Resource Survey found an average family income for the area of £138 a year, of which £16 was in cash. The best farms had less than £2 a week in 1965/6.[5] Below some minimum size of farm poverty is inevitable and another occupation must be sought.

The 'flight from the land' was a fact. The Republic had 38 per cent engaged in agriculture when the NFA was formed; 26 per cent in 1971. The flight from the land was halted for a few years by the prosperity of the 1970s but it is inevitable that farms which had once supported a family becaome too small to give full-time employment or to justify an acceptable modern income.

Increased production was rapidly driving Ireland in the same direction as New Zealand, towards 12,000 high-technology farms.

In the USA an adviser would not recommend establishment of a dairy farm with under 500 cows (RTÉ broadcast 25 February 1995). This would reduce Irish dairy farms to about 3,000. Electronically controlled cow byres require no farm labourers. This solution was rejected in Ireland, as in most European countries. No farming or political organisation could influence public opinion without the support of the small farmer. If the NFA was to have any impact it had to serve, and be seen to serve, the interests of the small farmer.

A transitional policy, actively retraining young country people for work in industry while giving income supplements to elderly farmers was discussed. The EEC Social Fund could be used to give education and skills to children of farmers and to attract industrial development for better jobs in farm areas. Rural society could not depend on farming alone.

The development of agri-tourism and other rural activities was the task of several IFA committees. Recent surveys show that over 40 per cent of farms have now nonfarm income. The science fiction of 1955 has become daily practice. A person in Belmullet can be part of a work team in Japan; the telephone booking service for Scandinavian hotels of an American chain can be made through an office in Ireland, as can passenger bookings for Korean Airlines. The secretary to a New York doctor lives in County Mayo. Many industries have become foot-loose and could, in principle, be located in rural Ireland.

Small Farm Organisation

The NFA wished to unite farmers. Small farmers lack the resources to run a powerful movement and the large farmers need numbers if they claim to be representative. In other countries small farm associations are withering as members leave the land. Membership in Denmark fell from 116,000 in 1946 to 20,000 (13,000 fee-paying) in 1991. Various small farm associations in Ireland have been attempted without success.

Little conflict arises between large and small farmers. Both stood to gain from better prices, better marketing. The NFA produced a number of Small Farm Plans. A Convention on the Small Farm was held in Athlone, January 1958, under the chairmanship of Myles Smith, vice-president for Ulster. A more elaborate proposal, was made to the Government in 1964. Its poor reception was an important cause of farm unrest. The ICMSA proposed a Two-Tier Milk Price to ensure small farms a share in subsidies. It was tried, but found unworkable. Direct acreage payments, proposed by the Germans and Swiss, were rejected by the NFA because it was supposed that tax payers in importing countries would never agree to acreage payments to farmers in exporting countries – a judgement proved wrong within the new structure of the European Union.

The Rural Development Section of the NFA produced a plan for small farm development in 1965. Though rejected by Government at the time, many of its proposals were incorporated in the government scheme of 1968.

Introduction of long term renting of land, enabling small farmers to expand without capital cost of land purchase, by Jim Gibbons as Minister for Agriculture was warmly welcomed (1977). The IFA press release claimed these measures had been the primary aim of the Rural Development Section for the previous two decades.[6]

It is inevitable that larger farmers are represented at National Council level. A one-man farm cannot afford the owner's absence; low-income farmers have no cars and public transport is ill fitted to attendance at late night meetings. Intelligence may be evenly spread in the population but education rises with the size of farm. Individuals beat the system only with difficulty.

Balancing Society

A prime social objective was to achieve negotiating strength to balance that of organised workers. This idea was developed by the Nobel Prize Economist Kenneth Galbraith. When any section of society becomes dominant it exploits

its strength. The exploited group builds a "countervailing power" to defend themselves.[7] The capitalists exploited the industrial workers; the trade unions balanced the capitalists' power. Trade unions in turn, having gained power. Laws to protect the poor became protection of the relatively well-off skilled worker against the community. One man with a strike placard can halt a vital industry. Farmers in 1955 tried to build a power base to gain the right to consultation and income parity.

The hard grind for the new Association was to put philosophy into practice.

21 The NFA and the UFU had close, friendly relations.
(l. to r. front) UFA delegate, R. Deasy (President, NFA), Joan Cummins (NFA), A. Algeo (UFU), J. Martin (President, UFU); (back) L. Smith, J. Greene (NFA), T. Cahill (NFA), J. O'Neill (NFU), J. Buttimer (NFA), S. Healy (NFA)

9

Relations with Government

Reform of agriculture required a fundamental change in Government attitudes, as was shown in Chapter 1. Decentralised ideas were expressed in the Recess Committee Report and in the Act establishing the Department; in the Policy report of 1922, and by the Commission on Vocational Organisation, 1943. The Senate was intended to be a vocational body representing trades and professions rather than political parties. In 1900, with no vocational structure of rural society in existence, it had proved difficult to give farm organisations effective representation as provided in the Act. After Independence government control was deliberately centralised.

In the post-war world the State became everyone's paymaster. Wages and prices were regulated; conditions of trade and employment were set; taxation grew from 29 per cent of national income to over 40 per cent. Prices for protected industries, industrial wages, particularly civil service pay, were fixed by Government without regard for the export market. The legal immunities for industrial action, conceded when workers were weak, were now exploited ruthlessly when they were strong.

Until the Trade Union Acts of the last century it was illegal for labourers to form any association agreeing to withhold their labour; similarly, it was a serious crime for other producers to withhold their produce from the market to raise prices by an 'agreement in restraint of trade'. Workers established the right to strike. The Act of 1906 protected them from claims for damages by people injured as a result of strikes. The right to form unions and to strike became accepted, even for wealthy or privileged employees. For the rest of the community the rigour of the law remained. The legal system gave a very uneven balance of bargaining power in Ireland.

The farmers' weakness in the market was recognised in other countries. For example, the Kapper/Volstead Acts in the USA allowed marketing co-operatives to form monopolies; the Marketing Acts in Britain, Northern Ireland, Scandinavia, gave farmer-controlled marketing boards monopoly powers to compensate for their weakness in the market. EEC encourages the setting up of regional commodity 'producer groups' provided they had monopoly status.

Irish farmers argued in vain that their produce price was their wage and that

22 The first NFA deputation to the Taoiseach at Leinster House.
(l. to r. front) A.M. Smith, J.N. Green, S. Healy, J. Buttimer, J. Callanin;
(back) G. Claxton, L. Smith, S.A. Sherry, S. Fitzmaurice, M. Ross.

their negotiating position should be as strong as a trade union's. Conflict was inevitable.

It might be supposed that farmers would dominate the state because of their numbers and because they provided 75 per cent of Ireland's exports. The contrary was the case. Farmers were so numerous that even a small subvention was an obvious burden on the tax payer. Food price increases were political dynamite. Subsidised produce became a gift to foreign consumers when exported. Industrial countries, with five per cent employment in agriculture could afford to redistribute income.

Farmers neglected to exert their numerical strength by voting to suit their economic interest. The paternal regime gave no way for the working farmer to influence decisions which governed his work and livelihood.

Centralisation of Government

Policy making in Ireland before entry into the European Community was an obscure process. The changing relations of the NFA with the Government may be a little more easily understood by standing back from the immediate events of the 1960s to set the scene in a wider context.

Ireland, as Sir Horace Plunkett observed, has an extraordinarily centralised system of government. William the Conqueror, knowing the confusion caused on the Continent by feudal lords like himself, established a strong central state in England in the 12th century. All swore loyalty and held their land directly from the king. London delegated little or no autonomy to Ireland lest local government should become a focus of national resistance. For the majority of the people there was no democratic structure of society.

Irish republicanism stemmed from the French Revolution. The extremist Jacobin Club had a strong influence on Wolfe Tone and the United Irishmen. The revolutionary government related directly with the citizen. It did not tolerate intermediate bodies. Indeed, as Karl Marx observed from history, revolutions lead easily into a dictatorship of the 'proletariat' in which the government party exercises centralised power in the name of the nation, allowing no intermediate organisation to develop lest it become a centre of opposition.

The new Irish Free State was naturally sensitive about its authority, but some deeper reason, such as the above, must underlie the quite abnormal political objection to farmers' involvement in milk prices, cattle, pig and turkey breeds or the marketing of their own produce.

Policy Formation

Independence was the prime concern of politics during the formative years of the state. From the Civil War came two main parties with no fundamental distinction between them. Social and economic policies were incidental.

Under the first Constitution of the Free State (1923) Ministers were individually appointed or removed by the members of the Dáil. Under the Constitution of 1937 the cabinet became a united body under the Taoiseach.[1] None the less the individual minister makes policy.[2] 'The policy of a government party, in effect, is the policy promoted by the cabinet minister with juristiction over the area concerned. A wise Irish voter wanting to find out about Fianna Fáil policy on third level education, will look, not at the party's official election manifesto, but at the policies which that Fianna Fáil minister promotes while in office.'[3]

'Formal responsibility for policy rests with the government, and responsibility for handling policy in particular areas lies with the minister. The civil servants, in principle, only carry out his decisions. Ministers come and go on the word of the Taoiseach or the electorate. The average life of a Dáil is about three years and there is often a reshuffle during that time. Consequently, the minister has usually lost office before his decisions bear fruit. One former official, who served with no fewer than ten ministers in his 20 years as Secretary of Posts & Telegraphs, described the ministers: 'for the most part, modest men who came without much in the way of ideas but who recognised that the best way to get results was to allow an old and well tried machine to function efficiently.' There being no system of Dáil committees like that in the USA, ministers are not interrogated on how decisions are reached. The minister bears full personal responsibility. In the Irish system, unlike that of the more advanced agricultural countries, the Ministers for Agriculture have very rarely had any practical knowledge of farming.'[4]

Relations with Ministers for Agriculture

Since ministers, in effect, make the Government's policy, the NFA had very close and frequent ministerial contact.

James Dillon spoke in favour of a non political union: 'If trade unions of banks, merchants, manufacturers and every body else are organising for the protection of their own specific interests then the farmers must also organise in a similar way if they are not to become the hewers of wood and the drawers of water for those who, in the last analysis live off the produce of the land. I want to see the day when the voice of the farmer will be heard in the land. I believe that if an organisation effectively to represent farmers is called into being by Macra the voice of the farmers will be neither mediocre nor shoddy in what it has to say.'[5] He studied and quoted the papal encyclicals on social organisation but in practice could not tolerate delegation of responsibility voluntary organisations – even to elect a board member on a state company or to regulate the breed of turkeys.

James Dillon, like his predecessor James Ryan, could only interpret disagreement in party political terms. Speaking at the NFA dinner he criticised NFA as 'a mouth piece of Fianna Fáil' and the *Journal* as a 'Fianna Fáil Rump' unrepresentative of the farming majority. He had used similar terms about the ICMSA: 'I am satisfied that the campaign at present being conducted by this association is a Fianna Fáil racket. I understand that the officers of this organisation are in touch with the Fianna Fáil Party for the purpose of organising a political campaign, allegedly on the subject of increased prices for creamery milk ... I state categorically now that I regard them as a tendentious political racket, that I do not intend to discuss with that association any matters relating to the creamery industry or anything else.'[6]

Similarly Dr Ryan, speaking in Wexford, accused NFA of favouring Fine Gael: 'Three years ago the Coalition reduced the price of wheat 63 pence a barrel. There was no NFA protest.' (This was not surprising. The reduction took place on the 9 October 1954 and NFA was launched on 6 January 1955.)[7]

Such politicians cannot understand, still less sympathise with, disagreement on agricultural policy or see an association as a struggle for a better livelihood.

James Dillon, as Minister for Agriculture, stated his theory of government to Louis Smith. It was wrong for government to delegate any state powers to a sectorial or trade association. In a democracy the people gave power to the Dáil;

the Dáil appointed the Minister, who was responsible for policy and and for his officials. The Government had no authority to delegate compulsory powers to voluntary bodies or to allow election to government boards which exercised powers of government – this in spite of the examples of British marketing boards or of continental farm unions. He could not agree that, since most voters cared nothing about the breed of cows, the more democratic and efficient system was to delegate minor regulation to the owners of the cows – subject to the State as guardian of the common interest.

A general election by all the citizens is a poor way of deciding whether a farmer may keep Landrace pigs.

Like his father, he firmly believed that people should stick to their own trades. Co-operatives were tolerable for processing of agricultural produce, but should not threaten the merchant. Basically the farmers' interest stopped at the farm gate and policy was for politicians; marketing for merchants.

Sean Moylan had the advantage of minimal formal schooling, leaving him open to new ideas. His appointment to Agriculture by Lemass was exceptional because he was given time to groom himself for the post while Minister for Education. His mind was uncluttered with misleading ideas. He selected a study-group of advisers who met with him in the Department of Education, helping to formulate policy.[8] Failing to be re-elected in 1957 he was appointed Minister for Agriculture from the Senate. He gave warm support to the foundation of cattle marts. He appointed Juan Greene to preside over the Advisory Committee on the Marketing of Agricultural Produce, the report of which did so much to change future government attitudes.

Moylan was cast in the Lemass mould. Had he lived even a year in office this history might be quite altered but his intolerance of inefficiency would have caused friction. He died after a few months in Agriculture.

Paddy Smith was most typical of the unpredictable and changing ministerial attitudes. He could be most supportive. At the NFA dinner, 1961, he made almost a recruiting speech: 'I believe there are too many separate organisations in the country. That is prejudicial to the farmers' interest. The existence of so many organisations is wasteful of time, energy and money. I cannot see why one organisation of a representative kind would not achieve not only as much but a lot more than a multiplicity of organisations dealing with the same problems. It weakens every organisation concerned. I know that some years ago a determined effort was made to solve this problem, but apparently without success. I want to say publicly and deliberately that it is a great pity that effort was not a success I hope another effort will be made.' He continued: 'The farmer is the loser ... I know it has been suggested that Governments cunningly connive at the situation on the principle of divide and conquer. I wish to make it clear that I would cheer loudly when I have to deal with one unified organisation.'[9]

In the NFA *Yearbook*, 1962, he wrote:

> I have often spoken of the desirability of having a strong and unified farmer representation for effective planning in agriculture and of the desirability of active co-operation between the agricultural organisations and Government Department. We are entering an era when such co-operation will be more essential than in the past and in the years ahead a well organised agricultural community will be a great asset both to the Government and to farmers themselves.

In December 1961, apparently referring to the NFA, he could say: 'A studied pressure campaign has been organised, with the usual weapons, misrepresentation of the facts, newspaper articles scurrelous resolutions and so on. I have a fair idea of those behind the campaign. The noise they make is disproportionate to their numbers, and the majority of farmers, who do not agree at all with their techniques can hardly be heard amid the noise these people are making.'[10]

Patrick Smith was a man of strong principles. Often in disagreement with the NFA, he was, as he said, Minister *for* Agriculture, a representative of the Government not of farmers. His duty was to the Government and to the public. This attitude, defensible though it may be, contrasted with Ministers of Agriculture in other countries. For example M. Pisani, French Gaullist Minister for Agriculture, 'orchestrated' farmer demonstrations timed to strengthen his bargaining power in the cabinet. He agreed that it was easier to turn demonstrations on than to end them. In Britain Reggie Dorman-Smith, president of the NFU, stated the farmers' case to the Minister one afternoon. Next morning he replied as Minister. Oddly, two Smiths, both from Cootehill in Cavan, were Ministers for Agriculture at the same time. They had little else in common.

Paddy Smith's fault lay in what he did not do. To his lasting credit he was almost unique in resigning on a matter of principle. The issue was the increases awarded to government employees, which he considered unjustified by the state of the economy and in view of low farm incomes. He misjudged the situation. By resignation he gave up the position of influence which he could use on the farmers' behalf.

He understood farmers. His departure was regretted in a formal statement by the NFA president.

Charles Haughey took office with expressions of good will on both sides. Expectations of an active policy were high, but he made no mention of farm organisations in his contributions to the NFA *Yearbooks* of 1965 or 1966. While personally friendly with leading farmers and with Paddy O'Keeffe, editor of the *Journal*, relations with the Association became uneasy and then hostile when, whether through lack of will or of influence within the Government, farmers criticised him for inaction. Disappointment sharpened the dispute. It has been said that 'the most critical time for bad governments is when they take their first steps towards reform.'[11] 'Lilies that fester smell far worse than weeds.'

The appointment of Neil Blaney as Minister reflected a government decision to deal strongly with the NFA. His refusal to meet the farmers was predictable. When Minister for Local Government, the Donegal executive of the NFA (his own constituents) had made repeated written requests for an interview to discuss drainage, without success, between 19 October 1962 and 15 October 1965.[12] To individuals he was freely available.

Even after the interventions of Sean Lemass and Jack Lynch had brought some agreement with the NFA, Blaney continued an aggressive policy. The Coakley and Gallagher explanation quoted above, seems to fit the events. Ministers made agricultural policy without detailed reference to the Cabinet or Taoiseach. However, once the Minister had taken a line of policy the Government felt forced to support him.

By 1974, when the *Yearbook* was revived, the climate had changed. Mark Clinton could emphasise marketing and good farming, with the feeling of a shared interest with regard to the European market.

Relations with Taoisigh

The NFA founders had grounds for belief that their contribution to policy making would be welcomed.

Eamon de Valera , in an election advertisement, stated that 'Fianna Fáil fully accepted' the responsibility for governing in accordance with the principles ennunciated in the encyclicals of Pope Pius XI on the 'Social Order'.[13] The encyclicals were not on theology but advice on social justice, as found in Denmark or France: 'The State should leave to smaller groups the settlement of business of minor importance, which otherwise would greatly distract it; it will carry out with greater freedom, power and success the tasks that belong to it alone, because it alone can accomplish these: directing, stimulating, restraining,as circumstances suggest and necessity demands. Let this system of subsidiary function be followed and a graded hierarchical order exist between various associations, the greater will be both the social authority and social efficiency...'[14]

In October 1956 the Taoiseach, John Costello, addressed an Inter-Party meeting:[15]

> There are six basic principles of Government policy. Number one, to favour investment in Agriculture above all other forms of investment ... the vital source and fundamental basis of whatever prosperity we have so far achieved or are in future likely to enjoy ... following our principle of co-operation the Government are anxious to secure closer consultation and co-operation between farmers themselves and the Minister for Agriculture, and growing participation by farmers' representatives both in the planning of policy and in the local administration of agricultural schemes. The Commission on Vocational Organisation made certain recommendations in regard to agricultural

> organisation in the parishes and counties and proposed the establishment of a National Agricultural Conference. In order to ensure full and frank discussion and consultation and co-operation between the Minister and his officials on the one hand and representatives of the farming community on the other the Government proposes to establish an Agricultural Production Council. The Minister proposes immediately to convene a committee of representatives of NFA, Macra, Muintir na Tire, ICMSA and IAOS. Having regard to the wide and varied interests represented by the NFA the Minister will ask that body to nominate four members and the other bodies two.

When the Committee met with James Dillon, Minister for Agriculture (12 October 1956), he suggested that the most practical approach would be that farmers should unite into a single representative organisation ... the Minister impressed on those present that he favoured this approach. Should it not be possible he would select organisations at his own discretion.

The organisations did not agree to unity. The opportunity to give the farmers an influential voice on policy was lost by inter-organisational squabbles. No blame can rest on the Minister or Government of the day.

The NFA dealt with de Valera as Taoiseach for a short period. He initiated the Annual Review of the agricultural situation.

Lemass introduced a new principle into government. Irish society was unduly biased to security – to survival in safe, pensionable jobs or ownership of property. 'The essence of Lemass's approach can be defined as the attempt to substitute the performance principle for the possession principle in Irish life.'[16] The search for security often developed into 'clientism' (the building of a web of favours to be repaid by neighbours or voters). Lemass fostered groups which could be made dynamic. In 1945 he tried to persuade the trade union movement to take an active part in development through the Irish Trade Union Congress: 'In a democratic state, the trade union movement must play an increasingly important part in national life, not merely as a guardian of the workers' interests, but as an essential part of the machinery of industrial organisation, accepting the responsibilities which relate to its real power, and proceeding from the stage of negotiating particular agreements with private employers to the stage of formulating and carrying into effect a general policy for the furtherance of the long-term interests of the workers as a class'.[17]

He extended this strategy to his relations with other business groups and to the formation of policies for economic expansion in the National Industrial and Economic Council (NIEC). It was natural to extend this partnership to farming as soon as a representative association of farmers came into being. Farm organisation disunity was apparently the root cause of farmers' exclusion from state consultative bodies.

It was equally natural that the 'grass roots' politician of the old style should oppose his policy. It struck at the base of local party structure of clientism, based on personal favours, family connections and loyalties, which gave security to politicians regardless of their contribution to national progress.

At all times Lemass made clear that the Government had final responsibility for policy – a matter that the farmers could hardly contest: 'The final responsibility for national policy – for determining the road of the nation's advance and organising the national effort – rests with the Government. It is our policy, however, to bring into active participation in the Leadership of this campaign the representatives of every organised economic element in the national community, and, to the maximum extent consistent with the realisation of the general policy, to leave the responsibility for the taking and executing of decisions with them.'[18]

The NFA took this as a declaration of policy, confirming the Government's acceptance of Social Encyclicals teaching that the central government should not be confused by undertaking to control every detail and the wisdom of devolving regulation of minor economic matters to minor, subsidiary organisations. It would give NFA a special place in consultation with the Government on all matters of agricultural policy, while not restricting Government consultation with any other body. There was a declared intention to devolve various functions to farmer control. The NFA understood that this would cover such questions as animal breeding and co-operative marketing.

There is no indication that Lemass revised that policy, though his priority given to agriculture declined as industry grew.

Civil Service

The structure in Ireland did not fit the professed ideology of the state. 'Ideally price protection should be enforced by the interested parties themselves, though supervision by the public authority cannot altogether be dispensed with. On this subject it must not be forgotten that the price of agricultural produce usually represents the reward of labour rather than the return on capital'.[19]

Under British rule few senior policy-making civil servants worked in Ireland. When Independence was gained most of these chose to work in Britain. The remaining staff of career civil servants filled the higher posts and carried on. For the most part they had joined the service straight from school by public examination. They were outstandingly honest and of high intelligence, but without experience or technical knowledge of markets or affairs outside the Service. They were administrators, not innovators or reformers. Their advice was a strong influence against any change in policy. The technical officers had usually little influence. They were 'always on tap; never on top'.

The National Archives covering the period to 1958 are now open to the public. One would expect to see revealed the development of Department thinking on the role of farm organisation when the NFA and other bodies were formed. In the slender file on the subject of farm organisations are two items casting a glimmer of light on their thinking. One is an unsigned, hand written note, from 1958, of guidance to the Minister on meeting an NFA delegation. NFA and ICMSA 'are not altogether friendly. The Minister for Agriculture has always had a little difficulty about representation. He could not deal with NFA above ICMSA. But NFA is widely representative and are on the whole representative of farmers. One of the reasons why the NFA has asked to see the Minister is probably because the ICMSA saw him.' The second is a set of newspaper cuttings describing farm unions in other countries. Underlined in red is a statement: 'The Netherlands Farmers' Union meets the Minister for Agriculture and his advisers once a month.' And 'The Farmers' Union has a special department for foreign affairs with 20 specialists', followed by a description of the International Federation of Agricultural Producers (IFAP).[20]

Early Government Recognition of the NFA

To influence policy it was necessary to establish the NFA as a responsible, respected organisation. Demonstrations and strikes might gain a price increase or an individual concession, as the ICMSA had done. Such strife would not achieve fundamental changes in policy. Dr Juan Greene, as president, was respected and not threatening. He was afterwards a member of the board of the Central Bank and several government and public companies. He believed in the natural goodness and rationality of man – that quiet reasoning could always convert those of another opinion.

Acceptance and co-operation of Government had to be gained at three levels: with the Taoiseach, the Minister for Agriculture and the Civil Service.

At first, particularly under Lemass and Moylan, the danger seemed to be that Dr Greene would lose the confidence of members by too close a connection with Government. Receiving confidential information he was unable to brief the Executive fully.

Appointed by Sean Moylan as chairman of the Committee on the Marketing of Agricultural Produce (1957) he influenced the future structures of marketing. He was less successful with the gamesmanship of petty politics, at national or organisational level. There the national objective was secondary and all criticism presumed to be politically motivated.

The first formal recognition of NFA as the representative of farmers was by Eamon de Valera as Taoiseach when he approached Dr Greene concerning protection for two farm machinery manufacturers in Wexford.[21] The Taoiseach agreed to an annual review of agricultural situation and policies. When drafting the press release of the meeting Jack Nagle, secretary of the Department, could not agree that the annual review had been instituted. On the insistence of Sean Healy the question was referred back to the Taoiseach for clarification. De Valera confirmed agreement on the institution of the Annual Review of Agriculture. It has been noted[22] that de Valera was always ambivalent on giving power to any institution outside the Government. He agreed the farmers' right to knowledge of policy and to influence policy by consultation. He was not delegating any executive function of government.

The attitude of Government to the Association is shown in the ministerial contributions to the NFA *Yearbooks*: Sean Lemass expressed good wishes in six issues from 1960 to 1966. In particular he referred to 'the delegation to representatives of producers and processors of a wide degree of responsibility for the actual sale of our products abroad' (1961).

In his Foreword to the 1966 *Yearbook* he wrote: 'Under the Second Programme for Economic Expansion we asked the farmers for a big increase in production, and I would like to take this opportunity to congratulate them for the truly magnificent way they are responding to the call ... It is the Government's aim to give every possible assistance to the farmers ...'

23 The NFA Council, 1956

Patrick Smith, as Minister for Agriculture, was fulsome in good wishes in six issues, 1958 to 1964. For example: 'In its four years' existence the NFA has had a fine record of service and achievement and has deservedly come to occupy a very significant place in our national life. All who are interested in the development of our national resources will be glad to see the Association growing from strength to strength.' 'Last year I commented on the excellent relations which existed between the NFA and my Department. I am very happy that these very friendly relations have continued during the past year and I look forward to close co-operation in the future in the firm conviction that by working together we can more readily achieve our common objectives of improving the lot of our rural population and of strengthening the nation's economy.'

Discussions took place on the Anglo-Irish Free Trade Area Agreement in 1965. Though the farmers were dissatisfied with the agricultural provisions, the NFA did not feel excluded from the decisions. This was in accordance with Lemass's pragmatic policy. To avoid confused signals and bickering between organisations he selected in each sector the most representative organisation for government consultation. In distribution, it was RGDATA; in industry, the Confederation of Irish Industry; in agriculture, NFA.

Charles Haughey's 'Message' in the 1965 and 1966 *Yearbooks* made no reference to the NFA function as an organisation. He was, however, the first Minister to address the National Council.

The 1969 *Yearbook* had no Ministerial message.

Annual Reviews

The Taoiseach, de Valera, and Smith, Minister for Agriculture duly met the NFA on 13 May 1958 for the first review of the situation. The NFA submitted a long memorandum on the state of agriculture. Great emphasis was placed by the farmer representatives on the need for research and information. Concern was expressed that the promised Research Institute for agriculture was not yet in place. The Danish farmers were criticising their government for an inadequate increase of £4.2 million for research in the same year.

Government was cutting expenditure on agriculture, reducing farm income while raising farm costs. The expenditure cuts had been made in the limestone subsidy (£175,000), land reclamation (£200,000) and in building grants.

Farm costs had risen in the year, partly due to government action. Increases were: Cost of living (13.4 per cent), Rates (30 per cent), cement (70 per cent), shipping charges had risen (£200,000). New industrial protection had increases machinery prices by £30,000; barbed wire was now 30 per cent above the price of Belgian imports; other protection was given to creamery cans, fertilisers, agricultural chemicals, building materials.

General wages had risen 20 per cent and agricultural wages 43 per cent while the Agricultural Price Index had risen 8 per cent in 1957 and would fall 2 per cent in 1958.

Marketing problems were discussed. Inadequate attention to the export market made expansion of production hazardous. For example, bacon exports had been allowed to fluctuate in a way that made Irish produce unwelcome on the British market. In 1954 12,700 tons were exported; in 1955 2,600 tons; in 1956 2,300 tons; in 1957 14,000 tons. Ireland had no agricultural attachés and the Department of Agriculture spent only £2,500 on market reports in the year.

The first Annual Review was completed in a manner satisfactory to both sides.

On succeeding as Taoiseach, Sean Lemass invited the NFA to discuss the economic situation on 27 July 1957.[23] Dr Greene suggested that the Annual Review should be modelled on that of Canada. It should be with the government rather than with one minister, so that the whole cabinet should be aware of the policy decisions

Discussion ranged widely, but dealt especially with export prospects and free trade associations. Market outlets would have to be developed if agricultural output was to expand in line with the Programme for Economic Expansion. For this improved quality and grading was essential.

Sean Lemass stated in the Dáil and confirmed in a memorandum to the NFA[24] that the Government recognised that the NFA 'are interested in all branches of agriculture and they welcome the prospect of regular and comprehensive discussions ... as well as their practical co-operation with Government in respect of specific aspects of agriculture ... The Government will arrange for an annual meeting with the representatives of the NFA to survey the general position, and the prospects for agriculture. It is the common desire of the Government and the NFA that these meetings should develop along the lines of a broadly based Annual Review of Agricultural Progress and Prospects.'

The position of the ICMSA was a main obstacle to full recognition of the NFA's position in representing farmers. Charles MacCarthy, president of the Congress of Irish Trade Unions, warmly supported NFA's participation in the various planning bodies. It seems that the inter-organisational split was the biggest, perhaps the only, obstacle to wider representation of farmers on government bodies.

Lemass saw a special role for the NFA as a representative of the Irish interest in non-governmental bodies, such as IFAP. He saw that it was in the national interest that finance be provided for this. When approached by Juan Greene he reiterated that the Government would 'continue to give financial support to the NFA in sending representatives to international conferences of farmers' organisations.' Correspondence on this is preserved in the National Archives. Between November 1957 and July 1959 half the NFA foreign travel expenses were reimbursed by the Government – a total of £1,391. Lemass wrote to Paddy Smith, Minister for Agriculture, asking him to give 'favourable consideration' to increased payments. As a result full payment was made for travel to the World Conference in New Delhi and Dubrovnik, and two thirds of Louis Smith's attendance in Rome. After the dispute of 1966 NFA never applied for any subventions.

In the Annual Reviews of farm affairs with the Department good relations prevailed. As late as February 1965 John Orpen, co-ordinator of the Review expressed to the National Executive his appreciation of the help of the Department officials in preparing his case.[25]

THE NFA AND POLITICS

The NFA did not owe allegiance to any political party. Irish parties had not been favourable, nor was there any party with which they could naturally ally. NFA followed the Macra practice, refusing to allow party politicians to hold office in the association. The prudence of such a rule was driven home when delegation members indicate loyalties with a clear, 'Yes, Chief' to Mr de Valera, or show the flag in a similar way to any other minister. Still, the NFA had to be involved with politics. The number of ways in which government action affected farming by restraint, tax or subsidy was constantly widening and the individual could do little to influence or inform the lawmakers except through association with his fellow farmers.

Juan Greene repeatedly encouraged farmers to be candidates in political elections when not holding NFA office. Past presidents of IFA have been particularly prominent in election to the European Parlement but less successful in Dáil elections.

Oireachtas Committee

The NFA Oireachtas Committee was formed in January 1961 in response to a request from the chairman of one of the political parties. The party wanted a regular briefing on NFA views on any legislation coming before the Dáil. Similar committees had been found necessary in other countries. One of the Committee's first actions was to visit the NFU in London to benefit from their experience. Rickard Deasy was appointed Chairman, to consult with the chairmen of any commodity committee affected.

Requests for briefing came from both sides of the House. Senator Owen Sheehy-Skeffington, representing Trinity College as a member of the unfashionable Left, offered to 'speak in favour of anything the NFA wanted stopped'.

Lobbying began at local constituency level. In rural areas the majority of voters were either farm families or merchants and providers of services to farmers. They also 'hung from the cow's tail.' It was rare for a TD or candidate to refuse to attend a farmers' meeting. All members of the Oireachtas were circularised by the NFA on important bills before the Dáil. For example, in 1961 all members were briefed on: the Poisons Bill, the Pigs and Bacon (amendment bill; the Agricultural Credit Bill; the Road Traffic Bill. Reference was made to the NFA in many Dáil speeches.

Farmers were encouraged to attend debates affecting agriculture. The Superintendent of the Oireachtas was co-operative in allocating seating accommodation for members in the Public Gallery. Sometimes the Minister, recognising the source of the arguments used by the opposition, would turn away from the House and address the public gallery where the farmers were observing the debate.

Politicians 'played to the gallery', knowing that their speeches and votes would be noted where votes counted.

Growing Discontent, 1964

Ministers for Agriculture, and to some extent their officials, did not willingly allow their sphere of influence, their power base, to be reduced. For example, Lemass, at Dr Greene's request, agreed to an appointment by NFA to the board of the Potato Marketing Company. The Minister's objection was not against the person but against the fact of election by growers.

As early as 1963 Paddy Smith was referring to 'campaigns of misrepresentation and false propaganda which have recently been instigated by certain interests. They appear to be somewhat crude attempts to imitate protest meetings and marches which have taken place in some Continental countries where radically different conditions prevail'. Rickard Deasy replied: 'It ill becomes an Irish Minister for Agriculture at this stage in the evolution of farmer/government relations all over Europe to use crude language and cruder methods in dealing with a voluntary, professional organisation, dedicated to the service of Irish Farmers.'[26]

The governmental principle of 'Cabinet Solidarity' obliged the Government collectively to support any Minister attacked. For a taoiseach to yield would involve a loss of 'face'. So, when the NFA demanded Paddy Smith's resignation because he insisted on political appointments to the Grain Board, refusing those elected by growers, it was explained by friends in Fianna Fáil that the NFA had made his removal impossible by demanding it.

Demands by the ICMSA for a milk price increase were not a threat to the political power base (though Dr Ryan had so considered them in the past). In fact such requests confirmed the client position of the farmers. When the NFA combined an attack on the inactivity of the Minister with an assertion of a right to consultation and to appoint the farmer representatives on marketing boards, a weak Minister felt threatened. The actual demands became irrelevant.

Greene stated at an NFA county dinner: 'No Government has as yet produced a coherent, comprehensive, planned agricultural policy for this country. Agriculture has had to develop on a day to day basis, not even being assured of stability for one year in advance. No industry can develop on these terms.'

Paddy Smith made a vigorous attack on NFA in December 1961, to which Juan Greene replied:

> To our Minister the NFA is a small pressure group, representative of nothing or nobody, or at best, it only represents a small group of of large inefficient wheat growers. Where he is concerned, you almost don't exist, nor does he or the Department wish you to exist. Whether we exist or not is not for me to say. I go no further than to say that this is the cross-roads for the NFA – you

either assert your existence or just pack up and go home and leave the fate of your industry to other and less knowledgeable hands.

It is quite evident from the speech of the Minister in Killarney that there exists a dangerously wide gap in the fundamental thinking over a wide and important area between himself and organised agriculture. In the first place he has denied them recognition as an authoritative opinion, or as having the right to express themselves as a representative body, presumably referring to the NFA in the following terms:

> A studied pressure campaign has been organised , with the usual weapons, misrepresentation of the facts, newspaper articles, scurrilous resolutions and so on. I have a fair idea of those behind the campaign. The noise they make is disproportionate to their numbers, and the vast majority of farmers, who do not agree at all with their techniques, can hardly be heard amid the noise these people are creating.

If the farmers are as ineffectual and miserable as the Minister believes, then ... it is the death knell of our whole agricultural industry ....

Ministerial kid gloves to soothe the brow seem to be reserved for the nonagricultural departments, while knuckle-dusters are part of the paraphenalia of the Department of Agriculture.

He is not the only one who persistently brands the NFA the enemy of the small farmer, probably because it is a good and useful way of of killing the NFA as the voice, even of the small man.[27]

The editor of the *Wicklow People* commented: 'Before relations between the Minister and the farmers become even worse there is a need for clear thinking and temperate talk on both sides. We can ill afford a state of war between any member of government and the men whose interests he is appointed to serve.'[28]

Juan Greene expressed his growing disappointment with results of the NFA approach in a speech at Timoleague: 'The experience of NFA over the last seven years has demonstrated that documented and supported arguments, together with patriotic restraint, are not paying dividends, and appear unlikely to do so in the future. Force has been shown to yield success after success for other sectors ... with see-saw effect on agriculture which went down while other sectors went up.'[29]

Rickard Deasy attempted to follow Greene's policy for the first two and a half years of his Presidency: 'Greene believed in the constructive approach – in reasoned argument rather than in force. But before he left office he had come to wonder if a change in policy was needed. I have resisted pressure to change course. Such a change would mean a reorientation of NFA policy. Demand from our counties now seems to dictate that we look at policy in this regard.'[30]

Deasy's more forthright statements in the same vein at the AGM of 1966 sparked off the major conflict with Charles Haughey. Yet Professor Seamus Sheehy could still write in 1968 on the need for agricultural planning at national level published by NESC: 'There is no shortage of platitudes concerning agriculture such as that it is a key industry in the Irish economy with considerable potential. What we are short of are the political will and the policies to realise the potential ...' The choice we see is clear: the country implements such a package or else it continues to speculate about potential while forced to bemoan the failure to realise it.' A sustained growth of six per cent a year was Sheehy's target.[31]

Disillusionment was growing among the members. In January 1962 the First National Convention was attended by 1,000 farmers. A motion was carried calling for more aggressive action. This was followed by protest meetings in the counties – 4,000 men in Wexford; over 3,000 marched in Clonmel; 6,000 in Mullingar; 8,000 in Tullamore; others in Lifford, Ballina, Nenagh, Navan, Naas, Ennis, Portlaoighse in the following months. In Dundalk 3,000 gathered in a snowstorm. The newspapers carried the stories and resolutions calling for more aggressive action.

The memorandum of Sean Lemass, considered the basis of NFA's relationship with Government, stated: 'It will be the ordinary practice of the Minister for Agriculture to inform the NFA about pending changes in the policy of the Department....' It appeared that the Fianna Fáil Government did a 'policy U-Turn' in Agriculture without giving any signals of their intention. The result was chaos and anger.

General wages continued to increase, by 30 per cent between 1962 and 1967; agricultural wages by 48 per cent. Further special increases of 12 per cent were conceded in the public sector. A 12 per cent increase in wages for agricultural workers was ordered in February 1968, without consulting the employing farmers.[32] By contrast, agricultural prices rose only 13 per cent. These wage increases had been achieved by trade union militancy – days lost in strikes, largely in the public sector, multiplied by seven between 1962 and 1966. As in 1956, cattle prices fell 28.8 per cent between Spring and Autumn.[33] This 1966 crisis had been clearly foretold by the Research Institute. The Government's planned increase in cattle output had taken place; the need for an active Meat Marketing Board to defuse the glut had been agreed, but no action taken. The farmers in the branches were very angry.

None the less, relations with the Government were still warm. The President was invited to meet the Minister for Agriculture for, 'a wide discussion on farm organisations, the Meat Board, Agricultural Prices, County Committees of Agriculture, the Advisory Services, NIEC.' The Minister for Agriculture spoke at AGM in July 1964 and to the Council on 21 December 1965. The Taoiseach agreed to speak at any celebrations that might be arranged for the NFA tenth birthday. John Orpen co-ordinating his eight committees for the Annual Review with the Department, expressed his thanks for the co-operation of the officials in the preparation of his case. The Association was satisfied with the 'Brown

Book' of the Department on 'Agriculture in the Second Programme for Economic Expansion'[34] and its forecast that 'average agricultural income per head (would rise) by 1970 to about 60 per cent above the 1960 level. The comparable increase expected in income per head for the economy as a whole is 40 per cent. Development in the non-agricultural sector will also help to solve the agricultural income problem through the creation of new employment possibilities for those who are under-employed in agriculture or who wish to transfer from farming to other occupations.'[35]

1965

Rickard Deasy's address to the AGM, 1965, was optimistic, constructive and friendly to the Department. He listed the flow of practical results from the Research Institute; changes in the Agricultural Credit Corporation; 'it would be a serious error for anyone to write off the considerable progress and success achieved by [Bord Bainne and the Pigs and Bacon Commission] especially in improving the quality and continuity of supply of our exports. We are now beginning to go about the business of exporting pig meat and dairy produce in a properly organised way.' He looked forward to the extension of centralised marketing to the meat industry. This would be essential if export expanded by the 50 per cent provided for in the Second Programme for Economic Expansion. The Knapp Report on the co-operatives set the scene for amalgamation into viable units and to the take-over of the Dairy Disposal Company.

Deasy hoped for co-operation: 'When announcing the decision to institute these Annual Reviews, the Taoiseach also acknowledged in a special way the position and wide interests of NFA. The Government is to take this into account in the appointment of farmer representatives to boards and committees dealing with agricultural matters. We are watching developments very closely in this connection, since they will reveal whether the Taoiseach's statement is to be developed, as I believe it was his intention, into the foundation for a fuller and more constructive degree of co-operation between the Association and successive Governments.'

The Minister, Charles Haughey's address to the Council (December 1965) could hardly have been more conciliatory: 'I am always – I would always be anxious to come and participate in your deliberations. Indeed I feel that as an ordinary member I am entitled to do so anyway.'[36] I accepted (your invitation) purely to indicate my good will and my desire and my intention to try and carry on the agricultural business of this country on a basis of co-operation and harmony with farmers and their organisations, and in particular with the NFA. I have said again and again that I as Minister and the Government of which I am a member are committed, firmly committed, to a policy of endeavouring to raise the standard of living of the farmer to somewhere in line with that of the urban community. Although it is a firm commitment on our part, we may not succeed in achieving it ... I promise we will work in the closest co-operation with your Association and with anybody else interested to try and achieve it.'

The Gathering Storm

The climate of relations had changed, however. A press release from a special NFA council meeting (26 May 1965) directed the president 'to seek an immediate meeting with the Minister for Agriculture in order to clarify the Government's new attitude to agriculture as revealed in the Budget'. While welcoming the new social welfare concessions for small holders, the council was adamant that the budget was a completely negative document as far as agriculture was concerned. 'In particular the small increase in milk price was given across the board, to dirty milk equally with clean.' The NFA had proposed that any increase should be used as a quality premium. Only by improved quality could new products and markets be developed.

Legality of Farm Protests

Remembering the experience of the milk producers and of their own Kilkenny members in the 1963 rates campaign, the NFA leaders were anxious to stay within the law. Two legal opinions were obtained. The first, by Mr Sean Breathnach SC, advised that, while the NFA could register as a trade union, this would offer no protection. The Trade Union Acts of 1871 and 1941 only gave protection in a dispute between workers and an employer. The Acts gave no protection to the self employed. The Government did not employ the farmers. Withholding farm produce would not constitute a trade dispute: 'The real danger the NFA could find themselves in is confederacy in restraint of trade. There is nothing wrong as I see it in protest marches provided the peace is not endangered or obstruction caused, but strikes or aggressive action if carried out by agreement could land them in trouble.'

He advised further on the Offences against the State Act, 1939, applying it to possible NFA action such as withholding part or full rates; withholding Land Annuities; obstructing roads, or other forms of militant action. Any attempt by the NFA to encourage its members to the above actions would render the Association an 'unlawful organisation' within the meaning of the Act and each member would be liable to prosecution.' Section 18 referred specifically to 'encouraging the attainment of a lawful object by violent means' and to an association which 'encourages or advocates ... the non-payment of moneys to the Central and or any other public body or the non-payment of local taxation shall be an unlawful organisation for the purpose of this Act'.[37]

Three years later, T.A. Finlay SC (nephew of Fr Tom Finlay SJ, the associate of Plunkett, and later Chief Justice), gave the same opinion.[38]

The forms of protest open to the farmers within the law were few. The Offences against the State Act, passed to enable the Government to protect itself against invasion or armed rebellion during the World War, could be used against civil, non-violent opposition to government. In practice it had already been used against the Leinster Milk Producers when withholding their milk supply. During the Kilkenny rates campaign, Neil Blaney, then Minister for Local Government, had threatened to have NFA declared an illegal organisation, making the members and staff subject to the same penalties as the IRA (28 November 1963).

It seems that the only people who can take concerted action against Government are state employees. (A major cause of the farmers' complaint was the Public Sector abuse of that power.) If the farmers would not use their votes in elections to influence economic policy they could be ignored with impunity. The forms of action taken in other countries were known, but illegal in Ireland. All hurt the protesters more than the Government or the public. They could only be adopted and supported under extreme stress. This came clearly through the Dáil Debate of 26 October 1966, referred to later.

In Spring 1965 the NFA undertook a 'go-slow' on payment of rates, withholding the first moiety. The action was widely supported because a substantial increase in taxes coincided with a reduction in farm incomes caused by the fall in cattle prices.[39] Next day there was a press release foreshadowing a commodity strike. On receiving a promise of compensation should farm income fall or non-agricultural wages rise by more than three per cent the National Council voted to call off the proposed action by 34 votes to 26.[40] Feelings were strong. The National Council debated a Westmeath County Executive's resolution calling on Rickard Deasy and others who voted for cessation of the rates action to resign.

Non-agricultural wages continued to rise. An NFA delegation went to the Taoiseach to 'collect payment' on the promises. The Department insisted that the meeting take place in Merrion Street, where there was an ICMSA picket. The ICMSA refused to join the NFA in the meeting if any topic other than the milk price was on the agenda. This posed a hard choice for the NFA. Support for the ICMSA picket would, as the NFA delegation understood the situation, mean forgoing benefits promised by the Government for all farmers. The Association officers had sharp disagreement in a meeting lasting several hours. Walking into the Department past the ICMSA pickets, Brendan Power, assistant secretary, NFA, remarked, 'That pulls the plug on NFA and farmer unity.' They passed pickets on the way in, but left through the unpicketed Dáil. Relations with ICMSA were permanently injured by this incident.

The NFA felt that the Government had reneged on their promise and deliberately played upon the division among farmers.

In August 1965 the President and John Orpen reported in detail to the Council on their meeting with the Ministers for Finance and Agriculture and Fisheries. The Minister for Finance did not understand the agricultural problems. Government departments were disregarding the undertakings given in the course of the

Annual Review. For example, information on details of water schemes, prepared in the Department five years earlier and promised five months ago, were still not available to the NFA.

'In the light of the reports given, the general opinion of Council was that the Association was now being obliged, through Government inaction, to take steps other than negotiatory ones to achieve justice for the farming community.'[41]

Questions to be asked were:

What practical steps the Government is going to take to help small farmers and when?

What action on rates and when?

What assistance is to be given to the dairy farmer to compensate for increased costs, what incentives to horticultural development, to cash crops, to increased pig production – and when?

There was frustration at the lack of progress in the formation of marketing boards and in devising incentives to quality production. Perhaps the biggest grievance was the 12 per cent pay increase given to the public service, with similar increases in other protected occupations, when the only concession on agriculture was in social welfare.

The only substantial policy innovation of the Minister was his instruction that agricultural advisers keep their distance from the farm organisations and the extension of further powers to the unrepresentative County Committees of Agriculture.

Reasoned argument could be ignored by the Government. The Association, remembering the jailing of milk producers and threats to their own members in Kilkenny, twice asked legal opinion on possible ways to put pressure on Government. They had two legal opinions that, 'while the NFA could register as a trade union, this would offer no protection. The Trade Union Acts 1871 and 1941 only give protection in a dispute between workers or between workers and an employer. The Act could give no protection to the self-employed. The Government did not employ the farmers. Withholding farm produce would not constitute a trade dispute.'

The Blood Committee, 1966

The hardening NFA attitude was crystallised by the report of the committee chaired by Bindon Blood in March.[42] Colonel Bindon Blood had one of the clearest minds on the National Council, where he was then Clare representative. He argued that

the success of NFA in real terms must be measured by the degree of support morally and financially which the ordinary farmer is prepared to give the Association. Moral and financial support from the ordinary farmer will enable the Association to do a good job for Irish agriculture in every field and the extent of this support will determine the bargaining power of NFA with the Government.

Therefore the policy of NFA must be geared in a way which will gain support from as many farmers as possible. While it is acknowledged that the Association must in the long term adopt a constructive and progressive policy, it may in the short term be well advised to follow a popular policy. It will be argued that NFA should not under any circumstances depart from a line which is obviously the most constructive. Such an argument, however, would be largely theoretical and would not be supported by practical considerations. It cannot be overlooked that in most cases farmer mentality and attitude is not as yet properly advanced to accept what is best. This whole problem stems largely around lack of education and appreciation of real values among farmers.

As a farmers' association the NFA must consider the circumstances as they exist and cannot base its approaches altogether on the ideal. The demands which NFA makes on behalf of Irish farming cannot be governed solely by economic factors.... Even at the expense of compromising itself to some extent, the NFA should be concerned primarily with its members' attitudes and regard its image outside farming as secondary ...

NFA rating as a farm organisation is today rightly at a high level. Unfortunately, however, its rating among its own members and prospective members leaves much to be desired ... The structure of the Association is not acknowledged by NFA members ... Farmers are indeed very often reluctant to acknowledge that they, as ordinary members of NFA, have a direct say in policy formation through channels which exist.

Farmers are more likely to rally behind and support a militant type of organisation than one which follows a more intelligent and constructive line. Organisational politics must therefore be considered, especially since the NFA has not a monopoly in the field of farmer organisation, particularly in the dairying world, which, as a matter of interest, constitutes an important basis of NFA income. While NFA should now be departing from the militant, largely negative line it must be remembered that farmers are slow people to change their habits and beliefs. Therefore, NFA rather than suddenly embarking on a non-militant type of approach must work towards a constructive policy on a gradual basis.

The sad lesson was learned that Trade Unions gained by strikes; the ICMSA picketed the Dáil and gained an increase in the price of milk; the NFA had reasoned a case and been ignored In many countries farm income had fallen about this time. In Denmark there was a commodity strike in 1961; Canadian farmers had a buying strike in 1974; French farmers were most demonstrative; even dynamiting a jail; Norfolk farmers marched on the British Parliament.

The press release from this meeting was headed 'NFA Prepares for Commodity Strikes' and set out the headings of farmers' grievances.

The evolution of the Association can be illustrated in the balance of staff. At the commencement there was a Secretary and an economist; then three economists and four organisers; in 1967 12 staff and one economist. (The NFU had 20 economists at this time.) The changed balance reflected the members' priority in the militant phase. If rational argument served no useful purpose with the Government, the farmers' present task was to re-establish the position of their Association. Dissatisfied with inactivity of the Government, the NFA officers felt forced by the members to become more like the ICMSA.

Annual Review, 1966

The depth of the discussions for the 1966 Annual Review can be judged by the the schedule of meetings. The Review was conducted in two stages, as planned at the preliminary meeting of NFA with the Minister and his officials [19 January 1966]. Stage 1 discussions lasted six days, on all topics covered by NFA committees. In Stage 2 the NFA Annual Review Team met the Minister for three days, covering points of principle unresolved in Stage 1.

Further meetings took place after the budget on 24 March and after the 'Mini-Budget' in May.

The Casus Belli

As is common in any conflict, the immediate cause of dispute was relatively trivial but reflected a serious matter.

When Charles Haughey, Minister for Agriculture, attended the AGM of the NFA in 1965, Rickard Deasy in his presidential address on farm policy said: 'failing a clear mind on the fundamentals of policy, no Minister for Agriculture in any country can hope to do more than proceed from one expedient to another in the mountain of day-to-day problems with which he is inevitably confronted.'

This statement was no more than the mild-mannered Juan Greene had said in March 1958, 'No Government has as yet produced a coherent, comprehensive, planned agricultural policy. Agricultural policy has developed on a day-to-day basis, not even stable for one year in advance. No industry can develop on these terms.'

At the next AGM (25 July 1966) Deasy launched a full attack: 'In the thirteen months past we have seen the Minister rushing from one expedient (whether political or economic) to another, while all the time shying away from key decisions on fundamental policy issues.'

He instanced several examples:

1) A plan for small farmers. Farm families were subsisting on incomes of £5 a week when a trade union leader complained that 150,000 industrial workers with £10 to £15 a week were below a living wage. There had been no serious effort to discuss the Small Farm Plan published by the NFA. 'How can anyone pretend that in regard to more than half our farmers we have heeded the fine words of 1916 and are now 'cherishing all the children of the Nation equally'?

2) Marketing. The NFA had published its proposals for meat marketing two and a half years before. 'Since last December the Minister has constantly been saying that he wished to have serious discussions with us on our proposals- indeed this was among the main reasons he urged on us to call off our rates protest – but he has so far not initiated those talks ... He has said publicly that he himself accepts in principle the need for a Meat Marketing Authority. In this he is in advance of his Department, which still blindly adheres to a dyed-in-the-wool belief in the Victorian doctrine of the divine right of the entrepreneur. Is it any wonder that in such conditions the great progressive efforts of our meat exporters are constantly frustrated by the opportunistic operations of the fringe?' The result was the failure of the Trade Agreement to deliver the price increase of £5 or £6 a head for cattle which the Minister promised. Instead we were getting 65 pence a cwt less than the British price for best cattle and the lighter cattle trailed miserably. If ever there was a case of a Minister trying to ride the whirlwind with reins of binder twine pieced together by slip-knots, this is surely it.

3) Delay in implementing cattle improvements, the recommendations of Marketing Committee Reports, the reconstitution of the Pigs and Bacon Commission and the Potato Marketing Company. The Department's determined efforts to head off the transfer of the Dairy Disposal company to co-operative ownership. 'How does the Minister, Mr Haughey, reconcile this with his statement in the Dáil on May 26th last – I am prepared to sit down and discuss any suggestion or proposal put to me by any deputy, farmer or organisation in a receptive way'. In January 1964 the Government itself concluded an agreement with NFA, an agreement which went part of the way to defining our status, but by its recent actions the Government has clearly broken the spirit of this agreement ... The lesson of recent months is that, as before, hand-outs will be made as, and if, political expediency dictates ...

Rickard Deasy reminded the farmers that after every war there has to be work of reconstruction: 'Farmers must now apparently once again gear themselves for renewed militancy, remembering always that militancy is a weapon not an end, that even during the fight, even if it is forced temporarily underground, constructive work must continue because in the ultimate the conference table is inevitable. When it is reached the militant who is a militant only and without long term constructive plans loses at the table what he has gained in his fight.

'If the Government then has decided that negotiations are to be fruitless, the

NFA will have to relearn and apply again in the coming months the only type of argument which the Government has shown itself capable of understanding, and to which it has shown itself ready to respond.'

The immediate causes of farmer unrest may be listed:

1) The collapse in cattle prices, following the increase in cattle numbers. The Anglo-Irish Agreement promised equal prices with Britain but beef was £2 a cwt lower in Ireland.

2) The publication of the Institute findings that farmers were making less than £5 a week – stated by politicians to be less than a living wage.

3) Civil Service demands for a rise of 12 per cent.

4) Increased farm taxation with rising Rates.

5) The weather. Rainfall 25 per cent above normal reduced yields, causing a fodder crisis.

More general causes were:

The contrast between the farmers' rights in society, as outlined in Mater et Magistra, and Government policy.

The failure of Government to take necessary action in marketing boards, prices or other practical improvements.

The general feeling that farmers were ignored on many technical aspects of farm policy, such as livestock breeds or transport.

APPENDIX

First Meeting with Mr Lemass

The role of the NFA in international affairs was stressed by Greene in his meeting with Lemass: 'We have no way of extracting money except by voluntary subscriptions and if for financial reasons we are limited in our endeavours to send delegates to these meetings we think it would have a serious adverse effect on the agricultural economy and on the country as a whole.'

Lemass replied: 'I think the basis of your freedom is your independence and financial resources should be got from the farmers themselves. You are better to have your freedom than depend on Government resources. I do think, however, that for instance the BGA were helped in collecting their money by legislation.'

Greene: 'We had thought that it should be possible for instance to introduce a similar system – say a levy on wheat – which could help finance our Association.'

Lemass: 'I agree one hundred per cent with the necessity for and always have agreed and fostered the maintenance of contact with other countries. I think it should be possible to help you along the line you mention.'

Minister for Agriculture: 'I don't know – how can you take wheat. I don't know Taoiseach.'

Lemass: 'In so far as international conferences are concerned I think we can help out there, where the official status of the international organisations arranging the conference is beyond question. In addition I could promise that the cost of engaging an expert to investigate any particular aspect of the agricultural economy would be financed and would be consistent with what applies to industry. We could not object to that'. (Transcript of meeting.)

His thinking was further elaborated with regard to farmers' representation on the NIEC (later the National Economic and Social Council, NESC) and in preparing the Second Programme for Economic Expansion: 'No adequate representation of farmers could be arranged which did not embrace all the interests. The addition of a considerable number of farmers' representatives to the Council would render it unwieldly in size, would greatly slow down its rate of decision ... That full consultation with farm interests is essential will not be contested, but, at least in the present state of affairs between farmers' organisations, this can best be achieved on a different basis. The arrangements for an annual review of conditions affecting Irish agriculture, which the Government have made with the NFA – while likely to be largely concerned with short-term problems and possibilities – indicate the procedure for consultation with farming interests which has been found to be most effective and useful, and which seems likely to be more attractive to farmers, organisations , and productive of results, than the membership of the NIEC. The Government have, however, made the farmers' representatives aware that they are ready to review this question of including farmers' representatives in NIEC if and when farming organisations ask us to do so.'

10

The Farmers' Rights Campaign

The Dispute of 1966

The government reaction was swift. During the AGM, when Deasy had finished speaking, Sean Healy received a phone call from the Department rescinding the promise of a milk price increase of 8 pence a gallon. At 7 p.m. another phone call cancelled the meeting arranged for the next day between the Department and the Livestock Committee, to be lead by Deasy. On the 26th Deasy, as president, telephoned the Minister about the meetings already arranged for the following week. He was told, 'There are none.' The conversation terminated by the Minister saying, 'I'll ring you.' There was no further communication by phone or writing.

The Department of Agriculture refused requests to meet NFA on the harvest situation, though the wet autumn of 1966 was causing congestion at the mills.[1] When the Grain Committee eventually arranged to meet officials of the Department the Minister cancelled his attendance on hearing that Deasy would be present. Such manoeuvres seemed intended to personalise the dispute and to distract attention from the real causes of unrest which Deasy had voiced in so pungent a manner.

The NFA continued to have fruitful discussions with other government agencies, such as Department of Industry and Commerce and the Valuations Office (of Finance), during 1967.

It seems that the Government misjudged the position, in spite of all the demonstrations, believing that Deasy spoke as an individual or for a small ambitious group. In fact he voiced an idea and a general discontent of farmers. Pressure of the ordinary farmer on the NFA leadership was extreme. Branches and county executives were close to revolt against what they considered to be a soft line at Headquarters.[2]

The NFA also misjudged the situation. They saw that the ICMSA had continued relations with the Government, under the same Minister, while breaking the law by picketing the Dáil. They concluded relations with Government would continue provided the NFA acted within the law. Efforts were made to stay within the law. There was nothing illegal in marching to Dublin, nor in waiting for an interview outside the Minister's office. That was not a picket. The machinery demonstrations were to be of vehicles legally licensed for the road

24 Farmers' rights dispute, 1966: two views of the Dublin march

and instructed to keep moving. Resulting traffic jams were not illegal. They were to be off the road by lighting-up time. There was to be no violence or abuse. The first clearly illegal activity was the withholding of tax payments. The Government's refusal to meet NFA could not be blamed on their illegal activity.

At a special council meeting with the Action Committee (8 September 1966) Rickard Deasy reported stalemate, without progress in farm supports or reorganisation of marketing. John Martin, president of UFU and Humphry Jameson, general secretary, addressed the Council, urging 'all farmers to make every effort to ensure that they retained control of their own business'. This was warmly applauded. Then Deasy and Martin, adjourned with Mr Williams, president of the English NFU.

Proposals for action were made. Withholding of Rates had been tried twice, with success in Kilkenny and less success a second time. It was felt that the Minister himself was organising this form of protest, since many farmers would be unable to meet the increased demands.

Few important decisions are reached unanimously. At each crisis – before passing the ICMSA pickets; marching to Dublin; machinery demonstrations; commodity strikes; withholding of tax – doubts were expressed and debates were sometimes long. Perhaps the Government would accept reasoned argument; perhaps promises would be kept; perhaps the NFA would alienate public opinion. Others believed that no argument would be heard without some kind of pressure. The National Council minutes record the expressed support from every county for any action taken by the President. Having reached that decision all stood by it.

The Presidential Committee (Jim Mullins, Kilkenny, chairman. Tom Finegan, Louth, vice chairman, Willie Bulfin, Offaly, with Brendan Power and Sean Healy) became special advisers on policy to the President. This group, popularly known as the 'Council of War', had no formal status and kept no minutes. In the course of dispute with the Government it was necessary to have rapid and confidential reactions. The National Council could not be called quickly; no group of a hundred people could be confidential. The president needed to be kept in contact with the counties and to have a small body with whom he could consult. Though influential, the committee was purely advisory. The Action Campaign was formally approved by the National Council (10 December 1966) and support was expressed for any action the president would take.

It was at this meeting that the 'fighting fund' was established. It might equally have been called 'contingency reserve' but the name expressed the feeling of the time. The bankers to the Association refused to approve an increase in the overdraft facilities. It was not considered a productive purpose.

Government Position

From the Dáil Debates (26 September 1966) it appears that there were two issues: 1) Fear of loss of face for the Government. 2) Whether the Government could devolve any functions to a vocational body to enable them to regulate their own affairs and giving the right to be consulted on agricultural policy. This had been the crucial issue between Dr Ryan and the Federated Farmers and again with James Dillon. NFA claims for direct election to marketing boards angered Smith. Consultation on such matters as livestock breeding and progeny testing became constitutional issues, undermining the structure of government, instead of routine arrangement of practical business.

The closeness of association between Government and farm association in other countries was well known to Irish farmers.[3]

Underlying, but impossible to discuss, was the growing belief of the Government that agriculture was not viable and so the future must rest with industry. If it was accepted that Britain, the only outlet for export, was becoming self-sufficient and that Irish agriculture would continue to stagnate, this position could be defended. The political weight of farmers could be ignored because the record of many elections showed that the farmer vote would not respond to economic pressures. This lesson was reinforced by the Waterford and Kerry by-elections of December 1967.

Progress of the Dispute

After cancellation of the livestock meeting the Department clarified the situation (27 August 1966): 'The Minister does not consider that, in the atmosphere created by the speech of the NFA President and his remarks at a subsequent press conference, a meeting with the NFA would serve any useful purpose.' To which Deasy replied that the Minister had his head in the clouds, 'It was time he came down off his high horse and visited a few sheep sales.' His casual words further inflamed the situation, being taken as a snide reference to the Minister's interest in horses and accounts of a riding accident.

Agricultural unrest was increased during the Autumn of 1966 by a sharp fall in incomes. The Institute published its estimate that 100,000 farmers had incomes below £5 a week. Labour leaders considered £10 a week less than a living wage. Some civil servants walking past the farmers into the Department had increases of £20 a week.

The NFA calculated from the National Farm Management survey that it required the income from 200 acres of feedling barley to pay the yearly wages of a docker and 100 acres to pay a member of the Garda, before pension and supplements.

The Institute researchers who forecast a glut in the livestock trade were criticised in the Dail as scaremongers upsetting the market. At first Sean Lemass accepted the official judgement that Britain could absorb the increased production through existing trade channels. The Anglo-Irish Free Trade Agreement was hailed by the Government as a diplomatic achievement giving Irish farmers equality with the British and worth £5 or £6 a head on cattle. The NFA did not read it that way. There was no undertaking by the British to buy Irish stores. There was no guarantee of price for the extra 200,000 Irish cattle available for export in response to Government's Programme for Economic Expansion. The NFA had pointed to the absence of a marketing policy. For nearly three years the demand for a Meat Marketing Board had been ignored. The fodder situation was weak because of a wet summer and and increased stocking. The British had no room for Irish stores. The effect was felt right through Irish farming. The Meath finishers were unable to move their stock; this reflected back along the market chain. Calves were bought for £4 in Blessington; one of the marchers from Bantry thought himself lucky to sell a six-month-old for £14 when the price had been £25 to £30 six months earlier. The fall in the price of the dairy farmers' calves was equivalent to a 10 pence fall in the price of milk. Tillage farmers were also hit by the wet harvest.

Government spokesmen urged farmers to hold stock for better prices. No indication was given of any steps to ensure credit availability for holding stock.

The NFA argued that, since the British farmer could not buy in Autumn, fat cattle should be sold quickly as beef. The *Farmers' Journal* was 'prepared to risk its reputation in saying forward cattle in good flesh' would make a good profit in Spring when bought at the current depressed prices.[4]

A record 630,000 cattle were exported live in 1967. This fulfilled the Government's target. Insufficient rail wagons were available and shipping facilities were short. The case for a Meat Marketing Board, long argued by the farmer organisations, could not have been better proved. No action was taken by the Minister to clear the backlog of cattle. The British Minister, Fred Peart, was more positive. He responded to pressure from the NFU and NFA by adjusting the Fatstock Guarantee Payments scheme to ease the market.

The NFA was acutely aware of the weakness in trade agreements between strong countries and weak. Almost the first action of Prime Minister Wilson on gaining office (16 October 1964) had been the imposition of a 'levy' of 15 per cent on all imports, including those from Ireland. This avoided the conditions in various treaties, which only covered 'tariffs'. The provision of an independent court of appeal in the Common Market treaties, giving security against such unilateral action, was one of the attractions to Irish membership.

Rates

In spite of falling farm incomes there was a sharp increase in taxation. Rates rose in every county, In the poorer Western counties the increase was over 10 per cent. In Mayo the rates rose from £4.1 per £PLV to £4.53; in Leitrim from £2.95 per £PLV up to £3.30. The more prosperous counties had lesser increases. The burden was resented because it was caused by increased salaries conceded to employees.[5]

It was an explosive situation.

RTÉ

When the Government issued a statement advising farmers to hold their cattle for a price rise, NFA issued a counter statement based on their assessment of the situation. The NFA statement was read on the Midday News (29 September 1966). The Minister telephoned RTÉ's Director General and the NFA comment was deleted from the Evening News. This caused uproar in the Dáil. Charles Haughey replied that he had not asked that the NFA report be censored. He had merely exercised his right to express his opinion. The National Union of Journalists were angered. The RTÉ Staff Manual laid down: 'It will be a fundamental principle of RTÉ policy ... that no pressure from any source whatsoever will be permitted to influence the content or presentation of its news bulletin.' The journalists sent a memorandum of protest to the the Director General, Mr McCourt. The NFA issued a further press release on the matter, which did not appear on RTÉ. As things turned out the NFA gloomy forecast was correct. This practice of indirect censorship was of long standing. By arrangement with the RTÉ journalists NFA learned to send its controversial releases just in time for the midday news, knowing that they might be suppressed in later broadcasts.

A main object of any demonstration is publicity. Newspapers gave full coverage. The national television and radio gave what the farmers and the reporters considered minimal comment. To gain television exposure the Donegal marchers spent an extra day crossing Northern Ireland, from whence the full BBC news coverage was received in the Republic.

The March, 8 October 1966

The idea that groups of farmers should march from all areas, converging on Merrion Street had been raised at the meeting on Small Farm Incomes. It was now discussed at length. The decision to march was reached in the light of correspondence from all counties expressing unqualified support for the President, offering active participation in any action which might be decided upon and confirmed by the National Council (10 December 1966).

There was hardship and loss for those involved. As TJ Maher said, the farmers could only get to Heaven through Purgatory – they had to fight for a permanent structure for their industry.

For seven weeks following the decision to stage a Farmers' Rights protest march organisation was intense. Statements clarifying the issues in dispute had to be drafted and publicised. Marchers, 10 to 15 in number, were selected from each county – men fit to walk 15 miles a day, of good national background.

Details were arranged of routes, with accommodation, food, back-up medical, publicity and other services.

Sixteen farmers, led by Rickard Deasy and blessed by both bishops, set off on the 217-mile walk from Bantry on 8 October. Similar groups from every county arrived together in Dublin.

The enthusiasm of farmers is hard to recall thirty years later. It is best illustrated by a few personal examples: Robert Wilson, Monaghan, aged 78, could not be restrained from marching; Jimmy Guest left his bride of one month to lead North Tipperary; Hugh Leddy was the largest small farmer, weighing 19 stone and supporting himself and family on 27 Leitrim acres. He was later jailed for traffic obstruction in the machinery demonstrations. Padraig Pearse, a teenager from Ballybunion, made the march as a last protest on his way to the boat for emigration; Harold Carter, Monasterevan, fed 150 marchers in his home. The livestock marts and co-operative creameries along the road provided food and shelter.

Children from rural schools turned out to cheer the marchers as they went by. Other farmers joined the main party to walk part of the way with them. Farmers from ICMSA and NFA were almost indistinguishable. It was estimated that 200,000 people were actively involved.

Eleven days later 400 marchers arrived in Dublin and were met by 25,000 – over 15 per cent of all farmers (Garda estimate). They marched three abreast from Parnell Square to the Department in support of their request that the Minister receive a deputation. News papers reported the greatest traffic jams seen in the city. As directed, the parade was in silence, there was to be no hint of violence; no unnecessary money was to be spent in the city.

The orders that violence should be avoided were constantly repeated, but sometimes the language used at meetings could be taken out of context as an incitement. For example: Tom Blake gave heated comment, 'as a member of Fianna Fáil for as long as I can remember, I am going to Merrion Street to clear the bloody lot out.' Or TJ Maher : 'We want to sweep Merrion Street clean. We are tired of a Civil Service that is trying to block every progressive development of the last couple of decades. It is wrong to blame Haughey entirely. He has advisers giving him wrong advice.'[6]

The IRA approached the NFA with an offer of support: 'You are interested in disruption, we are the experts.' Similar offers were made at local level. Various sources gave information that plans had been laid to use the NFA demonstration as cover for an attack and the burning of Government Buildings. For this reason the marchers did not seek to go to the Department Offices in Merrion Street, assembling for speeches outside the Mount Clare Hotel, in sight of Government Buildings. The NFA informed the authorities of these reports and kept close contact with the Gardai and the Army. This extended to the planning of the route and Army advice on the prudent length of daily journeys. To ensure that enthusiasm did not lead to any breach of the peace Willie Bulfin, national honorary secretary, was deputed to address the crowd. To diffuse the hotheads a meeting, attended by about 500, was held in the Teachers' Club hall, back in Parnell Square. There Fr Brady, of Roscommon, calmed feelings down. The precautions were necessary. In 1972 a march in protest at events in Derry on 'Bloody Sunday' ended aimlessly in Merrion Square, where a small violent group burned the British Embassy.

The effort of the marchers was equalled by that of their wives and neighbours, who carried on the farm work at home. When the leaders had to leave the sit-in to attend some meetings Sheila Deasy, Anne Gibbons and other wives sat knitting in their husbands' chairs.

The original plan had been to march to the Minister's office. When admitted the deputation would refuse to leave till satisfaction was received. There were enough farmers' sons in the Civil Service for the farmers to be informed of government plans. Locks had been changed and there was no chance of admittance. There was no thought of violent entry. The crowd stopped in Merrion Square to hear speeches from the leaders. Then, instead of entering Merrion Street, as expected and where Press and Television waited, the deputation turned left to Plunkett House, headquarters of the agricultural co-operatives, a few yards away in Merrion Square. The IAOS gave facilities for a press conference and toilet and other facilities for the duration of the dispute. This support showed some courage on the part of Paddy Kelly, secretary of an organisation heavily dependent on government subsidies. The owner of the Unicorn Restaurant, Mr Sidoli, himself of Italian farming stock, supplied hot meals free. The manageress of the Gresham Hotel, a farmer's daughter from Clare, provided hot water bottles for the cold nights.

While members of Macra were prominent supporters of the NFA demonstration, Macra, as an organisation, took no part. In response to a letter from NFA asking for their co operation in dealing with the RDS Spring Show, Hugh Ryan, president, told the Council that their work as an organisation was restricted to educational and social activity. As an individual he would give every help to the NFA when he had ceased to hold office in Macra. (In the 1980s he became deputy president of IFA.) The Council unanimously agreed. A few months later the same Council agreed with the incoming president, Joe Rea, that Macra take an agressive part in many matters of economic significance. This contrasted with the position of the IAOS, though equally dependent on government grant aid.

Attitude of Government Officials

The Gardai could not have been more helpful or courteous during the whole period of unrest. In preparation for the March on Dublin they discussed all routes and arrangements with the organisers. At Merrion Square there was no friction or effort by the farmers to pass the movable, three foot high barriers erected. The superintendent made it clear that, as far as he was concerned, no law was being broken by people waiting to meet the Minister. They were not forming a picket.

The farmers noted the contrast in attitude between the Department officials at local level (the county officers, CAOs and agricultural advisers) and the Dublin bureaucracy. Those who were in contact with production on the farm were with the farmers and helpful. In Meath the Garda superintendent, having inspected the traffic caused by the machinery demonstration, reported to Joe Bruton (father of the Taoiseach of 1994) that more machinery was needed at Stonyford Bridge. The Departmental instruction to Agricultural Advisers to report farmers activities to Dublin was considered harmful to their future co-operation in advisory work and an unfair pressure on them to act against their friends, relatives and their consciences. The Dublin civil servants, however, were interested in maintaining their detailed control.

On the Pavement

Nine men were selected to sit outside the Department: Rickard Deasy, TJ Maher, Michael Gibbons, Joe Dunphy, Bob Stack, Jim Bergin, Sean Holland, Tom Cahill and Hugh Leddy were told of the changed plan. Brendan Power, organiser of the demonstration, had chosen them with care; 'two big farmers; two small farmers; one hurler'. When the expected criticism was made in the Dáil or elsewhere that they were, 'Commandants, majors and colonels' or 'Blue-Shirts' it could be replied that none had served in foreign armies, three had held commissions in the Irish army and five had served in the Local Defence Force; Bob Stack had served in the War of Independence and had six All-Ireland football medals. Five – Tom Cahill, Joe Dunphy, Jim Bergin, Hugh Leddy and Bob Stack – were well known members of the main political parties; Michael Gibbons was brother of a government minister. Vice-presidents were not involved, lest all the leaders should be immobilised in jail. On leaving Plunkett House to take up his position outside the Department Rickard Deasy gave Sean Healy a letter addressed to his family, to be opened if the next stage of protest developed. The letter was returned unopened. A full hunger strike is a usual escalation in Irish protests.

The nine were to sit and sleep in the gutter outside the Minister's office, day and night, till he agreed to see them.

The Farmers' Five Demands were:

1) A national small farms development plan, together with a Western Economic Development Authority. [Such a western authority was still being requested by the Conference of bishops in 1994]

2) Effective floor prices for cattle and sheep, together with a farmer controlled Meat Marketing Board.

3) Restoration of dairy farmers' income.

4) General farmer control of marketing boards.

5) Effective general farm income policies, taking into account the impact of Rates and with special reference to tillage farmers.

Nights were cold on the pavement, so the farmers woke early to go across in ones and twos to wash and shave in Plunkett House. Breakfast, cooked by the NFA office staff and some well-wishers, was over about 8 o'clock leaving a quiet hour till the street woke up at 9 a.m. The civil servants passed by at 9.30. Forty or fifty supporting farmers joined the group at 9 and from that on there was a stream of visitors and bye-standers to reduce the boredom. Among the visitors expressing support for the delegation in Merrion Street was Maire MacSweeny, sister of the lord mayor of Cork who died on hunger strike during the war of Independence.

Minor events, like the meals sent, without charge, on heated trolleys from the Unicorn, broke the day. 'Survival suits', presented by the Canadian Federated Farmers, helped in the winter nights. At bedtime Hugh Leddy was heard to complain: 'Would you ever call TJ? He'll think of praying when we're all half asleep'. TJ was called away from the visitors to lead the Rosary and let the men settle down for the night.

Shortly after the farmers took up their vigil a group of workmen moved in beside them to dig up water-mains. Some doubt was felt on whether the mains needed repair, but there was no doubt about the discomfort of the pneumatic drills a few feet away. A phone call from Sean Healy to Charley MacCarthy, president of the Congress of Trade Unions, put a stop to the nuisance.

NFA House was burgled. No valuables were taken – there was no money on the premises. The safe in the basement was jemmied open and some hand-written notes were missing. Aware that any indiscretion in writing might lay the Association open to prosecution as an 'illegal organisation', minute books and other important documents were kept elsewhere. No one was ever charged by the Gardaí.

The NFA made efforts to remain within the law, Deasy asserting: 'We are not picketing the Department, only waiting to be received' (19 October 1966). And that: 'We would not picket Leinster House because we consider Parliament should be respected' (3 November 1967). Farm machinery was not to block the roads during demonstrations; essential services were to be facilitated; machinery should be home before dark for fear of lighting faults. The farm machinery must be legally on the road, moving from one place to another (7 November 1966). Payment of rates and land annuities were to be delayed, not refused. Neither Government nor local farmers paid attention to such fine points. The Action Campaign was formally approved by resolution of the National Council (10 December 1966).

Government Attitude

The Government's attitude was shown in the Dáil Debate of 26 September. Sean Lemass, as Taoiseach, had to support his Minister (and son-in-law): 'The Minister for Agriculture acted with the full support of the Government' in the dispute. 'To meet a deputation would imply victory for the NFA.'

He summarised the issues: 'As I understand it, an agitation is being made by the farmers' organisation for the purpose of forcing the Government to transfer to their leaders the responsibility for agricultural policy which the Government exercises. That is an impossible agitation for any Government to take cognizance of ...' He alleged that the NFA had premeditated strife. The cancelled meeting was: 'only a red herring. The real issue is whether the Government are going to remain responsible for agricultural policy or are going to abdicate that responsibility to the leaders of that organisation. We are not going to do that.'[7]

At times there was a denial that any problem existed. Lemass said: 'There is no difficulty if the farmers resume normal and reasonable discussion'. And, 'My advice to NFA is that they get up off the ground, walk over to their office and write a letter saying 'let us sit down and discuss'. If they do this discussions will take place.' Charles Haughey repeatedly asserted the same, 'There is no difficulty at all in the NFA coming to see me and discussing any aspect of agricultural policy. All that is necessary is for them to leave the steps of the Department' and make a request.[8] The meeting could take place 'within hours.' Such a letter from the NFA had in fact been delivered by hand to the Department of Agriculture before the sit-in began. Its rejection was the very cause at issue.

The Government attempted to minimise the conflict throughout the period, representing it as the action of a few malcontets. Neil Blaney could say in June 1968, 'relations with NFA are excellent'. At the same time Deputy Tom O'Higgins called him 'Minister for War'. In 1970 Blaney told the Dail: 'I have no dispute with NFA or any other farm organisation ... I would like to say to the deputies who are particularly concerned about NFA that there is no row outside this House.'[9]

The Minister explained that there had been no meetings with NFA in 1967 and 1968, only three in 1969 (compared with 150 in 1966) because NFA was not a member of National Agricultural Council. A letter from Jack Nagle, secretary of the Department of Agriculture, in reply to a request from Sean Healy, stated that there would be no more Annual Reviews with NFA. This would now be a function of the NAC. Healy replied that this was a clear breach of the 1964 agreement with the Lemass government – of which Jack Lynch, now Taoiseach, was part.

In fact discussions did take place. While the farmers sat in Merrion Street Tom Finegan and Sean Healy visited Haughey at his home without publicity. John Litton (Chairman of Macra in 1953 and then of the Agricultural Research Institute) was asked by outsiders to visit the Minister proposing peace. He was well received and came back with the same offer of a meeting as soon as the farmers left the pavement. The offer was rejected. By this time the NFA were so suspicious of the good faith of the Government and feelings were so strong in the country that they could not risk a gesture of trust. They considered that they had been manoeuvred into passing the ICMSA pickets with an unfulfilled government promise a few months earlier. The Association would dissolve if they were again deceived.

The NFA feared that they were being fobbed off as the ICMSA had been in 1959. On that occasion the ICMSA withdrew their pickets from the Dáil when given £0.08 a gallon of milk just before a closely run campaign for President of the Republic. De Valera won by only 120,000 votes. It was said that Charles Haughey, director of elections for Fianna Fáil, had bought the farmers' vote with an increase in the price of milk.

Several similar contacts were made. On one occasion the NFA deputation consented to meet the Minister, leaving their chairs outside on the pavement. On the 6.30 News from RTE next morning it was announced that the farmers had given in. Members of the deputation who had gone home were recalled at cock-crow from their first night in bed for many weeks and the vigil was resumed.

Truce

After 21 days and nights in the street there was a break. Sean Lemass, wishing to settle the dispute as almost his last act before retiring, sent a message to NFA giving, 'a solemn assurance of the goodwill of himself and the Minister in opening negotiations with the NFA'. A meeting was proposed (6 p.m., 8 November 1966). No preconditions were imposed. The Gardaí were withdrawn from the steps of the Department. As a gesture of goodwill, but with serious misgivings, NFA withdrew their delegation to the other side of Merrion Street and reduced their attendance to office hours.

Lemass had announced his retirement and Haughey was a possible successor. At the meeting with Lemass and Haughey, the Minister said that he expected to hold an unspecified but very important appointment in a few days. The following points were agreed with Lemass:

> There was need of a Meat Marketing Board and the Department of Agriculture was drafting headings for the bill to establish it.
>
> There would be full discussions on the Small Farm Scheme.
>
> NFA recommendations on remedies for the low price of livestock would be considered.

Pending an early meeting with the incoming Taoiseach and Minister, the truce was extended and a machinery demonstration was postponed. Since the issue in dispute was the failure of the Government to meet them, the NFA could hardly refuse to agree.[10]

The By-Elections

The presidential election of June 1966 had given de Valera an even closer margin over the Fine Gael candidate Tom O'Higgins. Now by-elections were due in Kerry and Waterford. The temporary thaw in relations, timed for electoral advantage, was effective.

The by-election campaign almost ignored the farm question. The opposition parties were confident of success in rural areas because of the low farm income, the price of cattle, the increased rates and the bad weather. Mark Clinton, shadow minister, unveiled Fine Gael's Agricultural Plan. The Department should be concerned only with policy. All operational functions should be exercised by a Rural Development Authority, which would also take in the Agricultural Credit Corporation and Land Commission. This authority would be controlled jointly with the farm organisations. This plan, launched during the by-election campaign (3 December 1966), attracted little attention. When Clinton became Minister in 1973 it was not resurrected.

The election result, with a very high turn-out of voters, was a major victory for the Government, winning both seats and gaining an overall Dáil majority. The Government increased its support in the rural areas. The *Irish Times* editorial commented: 'Even the Government must be surprised ... The Government success may show that support for the NFA is more apparent than real' (9 December 1966). If a few thousand farmers had shifted their votes the Government would have lost their majority in the Dáil, causing a general election. After the general election the ministers would have changed their attitude to the NFA and would probably have been different people.

The farmers vote had remained solidly loyal to party and family traditions, uninfluenced by the farm dispute. If the farm vote could neither be bought nor alienated the politician could ignore it.

Ministerial Change

After considerable delay, due to disunity in the party, Jack Lynch became Taoiseach, Haughey moved to Finance and Neil Blaney replaced him in Agriculture. As stated earlier, the new Minister was not favourable to farm organisation.

The NFA met the new ministers for three and a half hours on 21 November and again on the 25th. The delegation expressed themselves 'totally dissatisfied'. A reading of the transcript of the meeting gives a pathetic picture. There had been no deep discussion of agriculture at cabinet level. The two ministers disclaimed knowledge of the farming situation adequate to make any general policy decisions in the near future.

Small Farm Scheme

In November 1966 Charles Haughey had told the Dáil: 'I have prepared a comprehensive programme of small farm development which I will shortly be submitting to the Government.'[11] Assurance had been confirmed by Lemass (8 November 1966), giving Spring 1967 as target date for introduction of the scheme.[12]

It turned out that only the Pilot Area Scheme had been drafted. This was of no significance to the generality of farmers in any area. Only 5 per cent of the 106,000 farms under 50 acres in the 12 counties were involved. To illustrate the scale of the scheme: In Leitrim 302 farms with average income £246 a year would benefit; in Roscommon 372 farms averaging £236 a year; in Galway 267 farms averaging £232 a year. The increased in income expected for 1966 was 25 pence a week in Leitrim and £1 a week in Roscommon.[13]

Meat Marketing

The Meat Marketing Board proposals did not exist, in spite of the statement: 'Proposals for legislation for the establishment of a board in connection with meat export marketing have been in course of preparation and I shall bring them to the house as soon as possible.'[14] The new Minister said he would take several months to consult his advisers before he could form an opinion. The farmers were told, quite bluntly, that the new Ministers had been too busy with politics to heed the business of agriculture. Even the Meat Marketing Board, the necessity for which had been agreed by Lemass, and for which, according to statements at the preceding meeting, the headings of a bill had been drafted, had not been progressed. The Minister, none the less, claimed to be in full control of agriculture and full government support would go to him.

End of Truce

On 1 December the NFA declared that 'the truce is ended'. The delegation moved back across the street, placing four men at the Department of the Taoiseach and four outside Agriculture during office hours. Each county showed support by supplying 25 volunteers in turn on a roster.

The NFA felt they had been 'had' by the same ploy as that used on the ICMSA, without even gaining the 0.8 pence a gallon. This embittered the dispute.

Other Demonstrations

The next form of demonstration was by farm machinery parades. On 9th December 100 main roads were choked from 11 a.m. till 2 p.m. by slowly moving farm machinery. In Cork city there was a 'funeral of agriculture'. Parade marshals were instructed that there should be no violence. Vehicles should stay within the law, being fully licensed, always moving at their normal, slow speed. Essential services and funerals should be facilitated. The unwieldly machinery could not avoid disrupting traffic. Local farmers, not satisfied with the half measures of Headquarters, defied both Headquarters and Government. Some roads were fully blocked. There were complaints of minor violence in a few localities. In Athlone Charles Haughey's car was kicked; in New Ross protesters threw their placards at Paudge Brennan, parliamentary secretary to Local Government. In Waterford an ambulance was found to be loaded with barbed wire in an effort to circumvent (or perhaps to ridicule) the obstruction.

1967

Farm machinery was out again on 9 January 1967, in spite of Government warnings. It was estimated that 75,000 farmers and 15,000 farm vehicles were involved.

The Donegal Executive, by contrast, did not participate, deciding that the obstruction of traffic would be counter-productive, alienating the sympathy of non farmers.

The obstruction of traffic by the machinery demonstrations of January caused inconvenience to the public and some real grounds for complaint. The Government Information Bureau, silent on inconvenience to the public when a few relatively well paid trade unionists closed banks, schools or public transport, strongly denounced the inconvenience. A particular allegation was that Mother Mary Martin, foundress of the Medical Missionaries of Mary and a popular heroine, had been delayed and abused by farmers as she drove to the Drogheda hospital with an urgent supply of blood plasma. She issued a statement that she had been given every facility to pass; 'we met with every courtesy from the farmers ... They lifted tractors and other machinery out of the way to get us through as soon as possible ...' there were no urgent medical supplies involved.

A circular from the Amalgamated Transport and General Workers' Union instructed its members that removal of farm vehicles was not part of the duty of local authority workers.

Three hundred university students of agriculture from UCD marched from Mountjoy Jail to Merrion Street to protest against the jailing of farmers (11 March 1967).

Farmers announced a ban on hunts if any member of the Government was riding. Charles Haughey was active in fox hunting.

Farm women drove their tractors from Earlsfort Terrace to Merrion Square and O'Connell Street, stopping occasionally to give away vegetables from their trailers and to tell the Dublin housewives of their case. Other women marched from O'Connell Street to the Department.

Rates Campaign and Prison Sentences

The dispute escalated on 16 January, when the National Council advised farmers to give priority to payment of merchants' bills over payment of rates, land annuities and tractor tax. This was familiar territory for the Government and their response was well rehearsed from similar incidents in the 1930s, 1940s and 1950s. Five days later five farmers were charged in Tipperary Court. On 13 January, 18 Tipperarymen and 16 in Drumshanbo were arrested for picketing the rates offices; 10 Waterfordmen were charged for traffic obstruction. In all about 200 were sentenced to a fine of £6 or eight weeks in jail. No fines were paid. Farmers refused to sign bail-bonds, as that would be considered an admission of guilt. This 'criminal record' would debar them from travel visas for the USA and other countries.

To put the dispute in perspective it must be remembered that the Rates against which the small farmers of Leitrim protested at Mohill, for which they were jailed and for which farmers' property was seized by armed force in 1970, were abolished in 1982 because the Supreme Court found them to be unconstitutional and 'The whole system is shot through with unecessary anomalies and inconsistancies.' The Government, it was said, had 'made a mountain out of a Mohill.'

The Government placed full page advertisements in the papers, showing an extreme version of farm subsidies, implying that the farmers collected £56 million in subsidies and paid little tax. The subsidies listed contained non-agricultural items and did not allow for local taxation paid by farmers. John Orpen recalculated the figures, showing a more realistic £13 million in agricultural subsidies and that subsidies to industry were substantially greater.

Government aid to agriculture had been a controversial issue from the first years of NFA. The official figures included non agricultural items, such as the Botanical Gardens in Dublin, subsidies to fertilisers which were for the benefit of manufacture, education – whereas industrial technical education was placed under the Education vote. Relief of rates on small farms was counted as a substantial subsidy, while Earned Income Relief and other, much greater, remissions of income tax were ignored. It had been agreed at an earlier Annual Review that these contentious figures would not be used.

Life in Jail

Enforced idleness in prison was a new experience which brought men together. The prisoners organised to gain humane treatment. The older men had reason for concern that the younger members be corrupted and sexually harassed by criminals. A disused wing was made available, provided the farmers would clean it for themselves. Prison became a training school for future leaders such as IFA President Alan Gillis. The minutes of the 'NFA County Kildare, Mountjoy Prison Executive' record meetings in Cell 37, starting on 25 February. In this forum future action was debated. Some felt that, having been jailed for a petty fine, there was little more could be done to them. A few farmers with their chain saws could fell enough trees to immobilise the country. Wiser heads pointed out that farmers, as settled members of society, would suffer most from this disruption. The practical decision was taken that, since a milk strike was no longer effective when dried milk was available, the Leinster Milk suppliers should commission Arthur Andersen, international consultants, to advise them on the co-operative takeover of the pasteurising and distribution of the Dublin milk supply (completed in 1985, with the takeover of Premier Dairies by Tir Laigheann Co-operative). It also speeded the integration of the LMP within the IFA.

Commodity Strike

The next escalation of the dispute was the general commodity strike (6 March 1967). Sales of meat and vegetables were halted for a week. A test of strength came at Donegal Fair. The Minister exerted his influence in his own county, but at the end of the day the commodity strike had held. A telegramme was sent to Headquarters, 'Donegal. 6 p.m. Not a beast in sight, not even the Minister.' Farmers were charged with intimidation because they had approached the few farmers coming into the town, explaining to them what the dispute was about. The court acquitted the farmers.

There were no cattle and no auctioneers at the Dublin Cattle Market, which was patrolled by 400 farmers. On the same day the Taoiseach, Jack Lynch, offered leniency in treatment of prison sentences, fines and future prosecutions if the NFA would call off illegal actions.

The NFA needed support from any source available. International fraternal messages were received from farm organisations including Belgium, France, Germany, Norway, Denmark, Switzerland, Canada, Israel, India, Italy, Japan, Australia, New Zealand.

The Church and Farmers' Rights

NFA was particularly careful to gain clerical approval. They needed moral authority endorsing their actions. The papal encyclical *Mater et Magistra* was practically a handbook of the Association. Rickard Deasy, Micheal-Owen Fogarty, Tom Cahill and Sean Healy had discussed it with the Pope in Rome. Rickard Deasy read to the marchers in Merrion Square a letter from Mgr d'Ascenzie, speaking for the Holy Father, expressing support, 'for the efforts of small Irish farmers to get a fair return for their labours'. Mgr Legutti, Vatican representative at FAO, also rang the Nuncio in Dublin giving the Vatican view. Setting out from Bantry the farm marchers, led by Deasy, were blessed on their way by both Catholic and Church of Ireland bishops; the Monaghan marchers were seen off from Castleblaney by Dr O'Callaghan, Bishop of Clougher (who said he would have marched to Dublin with the farmers had he been fit), and the Presbyterian Moderator.

During the sit-in Rickard Deasy received a letter from Mgr d'Ascenzi: 'From the documents I have been able to know and appreciate more than before the lively and well-balanced trade-union action the NFA is carrying on in favour of Irish farmers. I feel it a duty, moreover, to offer a word of appreciation regarding the content of the Declaration (the nine-point Declaration of Farmers' Rights, which Deasy had sent him: see page 152) which is in complete conformity with the fundamental values of the human person which are stressed by the teaching of the Church' (26 October 1966).

The organisers repeatedly asserted the issue was not a penny or two on milk price. While income was involved, the basic issue was the right of farmers to have a say in the conduct of their business. The cumulative errors made by the Department of Agriculture over many years made this urgent. It was this issue that brought clergy of every denomination to the farmers' side.

However, the climate changed. A moral issue was raised by the Bergamo Association of priests (a group of junior curates whose parish priests would not allow them to take active part in economic issues; they met regularly in NFA House; Bergamo was the birth-place of Pope John XXIII) and by the Church in Ireland of the Methodists when a commodity strike was mooted. The half-yearly meeting of the bishops in Maynooth College considered the farm protest as civil disorder and the proposal to withold essential farm produce from market as failing in consideration for the rights of others. It was proposed that the NFA action be condemned and that clerical support should cease.

Archbishop Morris, of Cashel, and Bishop Harty, of Killaloe (brother of the man who brought the pigs to Korea: see chapter 3) were concerned. They asked that action be postponed till the next meeting. Dr Morris visited Sean Healy in Dublin informing him of the situation. From this it was arranged that Sean Healy meet Cardinal Conway in Armagh. This was a six-hour session at which the Cardinal accused the farmers of being 'off the rails' and in conflict with the duly constituted authority of the State.

The NFA leaders were very much concerned that a public shift from open support for their actions should divide the loyalty of their supporters, particularly among non-farmers. They saw the change as a Church response to Government pressure.

Members of the Executive of NFA, TJ Maher, Michael Gibbons, John Richards Orpen, Tom Cahill, Tom Finegan, Tom Blake, Willie Bulfin, Jim Mullins and Sean Healy went to Armagh. Michael Gibbons was an outspoken man. He pointed out that no trade union had been denounced by the clergy, even when grave harm had been caused to the public by relatively well off groups. The NFA had never questioned the legitimate authority of the State. The NFA instructions had consistently ordered restraint – that Ministers be received in silence. He reminded the Cardinal how, at the foundation of the state, the Church sided against those seeking a break from Britain – a mistake which should never be forgotten. The same mistake should not be repeated through over-respect for authority. Michael Gibbons' uncle, Fr Michael, obtained his doctorate from

Fribourg with a thesis on the state and tyranny which prepared him for this debate.

No more was heard of the matter.[15]

The National Agricultural Council

The Government instituted the National Agricultural Council (NAC) in an attempt to fill the vacuum caused by its refusal to meet the NFA. Neil Blaney announced its formation on 9 February 1967, The Department invited a number of rural organisations to nominate members to the Council. Little regard was paid to the size of the organisations or their functions. NFA and ICMSA would have two members each. The politically controlled County Committees of Agriculture and the BGA would also appoint two members each. The Minister would appoint six members himself. To clinch political control the Minister would be Chairman and the Department would provide secretarial services. The same ploy had been used in the first proposals for the board of the Research Institute. By giving equal representation to many small associations the larger farm organisations would be without influence. Past experience of 'consultative councils', which hardly ever met and had no influence, inspired no confidence. (See chapter 1.)

The NFA National Council rejected the proposal (23 February 1967). In principle such a body could have been useful. NFA had co-operated with James Dillon's shortlived Agricultural Consultative Council. The proposed NAC had fundamental flaws. In the NFA view no organisation could represent farmers if farmers did not elect their own representatives; it was absurd for the Minister to claim chairmanship of NAC if it was to negotiate with himself; there was no provision for a secretariat or for the calling of meetings except by the Department; there was no clear statement of its functions.. The Minister refused to discuss the functions of the Council with the NFA. NFA requests for information on NAC proposals for the wheat harvest failed to obtain an answer. At this meeting the NFA decided on a passive non-co-operation with all government services and agencies. The intended role for the NAC as a representative body became clearer when the Minister invited NAC to represent farmers on the National Industrial and Economic Council. Sean Lemass had indicated that the NFA should fill such a a representative function. A letter from Jack Nagle, Secretary of the Department, to Sean Healy stated clearly that there would be no more Annual Reviews.[16]

Dr Garret FitzGerald, no particular friend of the Association, wrote at the time: 'this is clearly an anomalous situation, which derives from the fragmentation of the farming community in Ireland. Although the NFA is by far the largest and the only properly staffed farmers' organisation in Ireland, a number of other bodies challenge its right to represent the farming interest. Its principal opponent is the ICMSA, with which it wages seemingly perpetual guerrilla warfare ...

There does not seem to be any reason why agricultural interests should not be represented on the NIEC in the same way as management and union interests in the non-agricultural sector, that is by direct nomination by the relevant bodies.'[17]

Glimmers of Peace

The budget of 1967 gave concessions to agriculture. Small farms were relieved of rates on the first £20 of poor law valuation, with a reduced rate for a further £10 PLV; £450,000 was allocated for a small farm development and pilot scheme. Milk price was raised £0.05 a gallon. The total came to £5 million.

Various contacts were made behind the scenes, as a result of which Deasy announced that NFA would not take any further action outside the law and was prepared 'to let by-gones be by-gones' (13 March 1967). The following day the Government expressed willingness to start talks between Minister and NFA. Sixty-five farmers were released from Mountjoy and Limerick jails.

The NFA Council was due to meet on 28 April. While the NFA would have preferred that the budget concessions be given for productive development rather than as hand-outs, the climate had improved. All indications were that a decision calling off the 'go slow' on rates payment, at least till discussions took place, would have been reached.

Next week, however, a further two Clare farmers were jailed for refusing to sign 'bail bonds' concerning actions taken during the commodity strike. The renewed imprisonments came the day before the NFA met the Taoiseach and Minister for a seven and a half hour meeting and helped make it sterile (21 April 1967). Kevin Boland, Minister for Local Government, threatened prosecution and property seizure for non-payment of rates (14 April 1967). On the 24th the Taoiseach made an unprecedented television broadcast warning of further imprisonments.

Seizures

Five days before the scheduled National Council meeting, the Government took action.

The NFA knew that property would be seized for non-payment of rates and the names of the prominent farmers targeted had been given to NFA by friendly civil servants. Michael Gibbons had served as a captain in the Irish army. He represented Kilkenny on the National Council and had been chairman of the National Livestock Committee. Later he was president of the co-operatives. (IAOS). His brother, Jim, was parliamentary secretary to the Minister for Finance and afterwards Minister for Agriculture. Though not himself involved in party politics he had been approached by Fianna Fáil to stand for election with a view to becoming Minister for Agriculture. The family of Mrs Anne Gibbons was prominent in politics in the Fine Gael party. She was chairman of the NFA Farm Family Committee and was later on the National Council. Jim Mullins was also from a Fine Gael family. He was on the National Council and chairman of the Grain Committee. Michael Dillon, a prominent sheep farmer from nearby Tullarone (not the journalist), was the third farmer involved.

Plans had been laid for the neighbours to rally round when the bailiffs came. The timing of the seizures was not known, so it was a surprise when Mrs Gibbons looked out at 6 a.m. and found the house surrounded. The 'Black Maria' was at the door. A large force of Gardaí and soldiers backed the State Solicitor and Sheriff. The children were wakened by the Gardaí walking all through the house. There seemed no chance to contact the neighbours. The phone was not accessible. The family, including the children, were imprisoned in the house. (The legal position of the state in refusing to allow the children to go to school does not seem to have been examined.)

Michael Gibbons played for time till 8.30, by which time neighbours had gathered.

Television and the other media were present in force when household goods were taken out to the lorries. Considerable play was made of luxury items – the dish-washer, which was hardly remarkable in a household with 13 children; the 'horse-box', also highlighted, was in fact used for carrying cattle.

A similar raid was made on Jim Mullins, where household goods were seized, and on Michael Dillon, whose sheep were taken.

The loaded lorries were followed by a cavalcade of farmers to Island Bridge barracks in Dublin, where the farmers kept vigil to report any effort to sell the goods. In due course the property was returned, unharmed, after the dispute.

Publicity was the major objective of the State. The message was that prosperous farmers were refusing to pay taxes while all other citizens were burdened. Viewers were not told of family circumstances or the cause of dispute. This was effective Government propaganda. By showing the lifestyle of a couple of progressive farmers the impression was given that all farmers were similarly prosperous. No distinction was made between evasion of taxes and a temporary delay in the normal payment of taxes used as a means of publicising a dispute.

Public Sympathy Was at Stake

It was common knowledge that the cabinet ministers were at loggerheads. Neil Blaney revealed the situation years later in an exclusive interview with the *Sunday Independent* (4 March 1973). Describing Jack Lynch as a 'political non-entity' incapable of leadership and decision-making, he went on 'Jack Lynch was never a friend of mine, I would not even regard him as a working colleague. From the first day we did not see eye to eye'. His disillusion with Lynch dated from the 1951–4 period. The interpretation of the Government's contradictory actions was that the Taoiseach wanted to calm the situation, but certain hard-liners among the ministers disagreed. Discontented with the Budget and other concessions to farmers, they wished to forestall any settlement and force the NFA into NAC. The NFA expressed this suspicion with a headline in their page of the *Irish Farmers' Journal*: 'FIVE DAYS TO PEACE AND THEN....'

Further Imprisonments

Nineteen farmers were imprisoned for refusing to pay traffic fines arising from the machinery demonstrations. By mid May a hundred farmers were in jail.

James Flood, of Maynooth, went on hunger strike. He would not sign a bail bond which 'gives farmers the stigma of criminals' (9 May 1967). The significance of such a stigma was increased by the USA visa regulations. Those with a criminal record would be refused entry. Frank Monohan, of Straffan, was father of five and certified by his doctor as unfit for prison. The Executive directed him to sign the bond rather than endanger his health. Never the less he presented himself at the prison, asking for admission so that he could join his comrades and was frustrated when the prison doctor also certified him unfit to serve a jail sentence. Cyril Booth, substantial farmer in weight and otherwise, was angered when someone paid his £5 fine without his permission. He refused to leave prison. The warders dragged him outside the gates, but there he sat demanding readmission. The unusual request of Bill Diamond, Athy, the tallest demonstrator – two metres tall as he said himself – was conceded. He was given a saw and tools to lengthen his bed in Portlaoise. Rickard Deasy and Sean Healy managed to be inside Mountjoy to give encouragement to prisoners on their transfer from Portlaoise.

In March 1967 Rickard Deasy reported to the National Council that: 'a firm message has been sent out to us from the prisoners in Mountjoy and Limerick jails. They told us in no uncertain terms that under no circumstances are we to make a soft settlement for the sake of getting them out of jail quickly.'

Niggling

To increase pressure on NFA supporters many minor means were used. The normal procedure to take against those defaulting in payment of rates or other taxes was by summons in the District Court, a cheap and simple remedy. Civil bills were now issued from the Circuit Court. This formality cost about five times as much, charged against the defendants. It also enabled a judgement mortgage to be registered against the land so that it could be sold in satisfaction of the debt (March 1967).

The prosecuting lawyers explained in court that they had been instructed to apply for the taking away of driving licences from farmers involved in demonstrations. This new procedure in the case of traffic offences was a severe penalty on farmers, who needed cars and tractors for their day to day work.[18]

In Leitrim the Gardaí were instructed to prosecute the small farmers accused of picketing the rates office under the Anti-Trade Union Act of 1875. This British act for the suppression of poorly paid factory workers had been reversed to give trade unions immunity from prosecution for pickets and strikes. The Act in question does not appear to have been used in other counties. The fact that it was invoked against farmers with incomes below the poverty line caused deep resentment,[19] especially as the tax was considered unjust – an opinion upheld by the Supreme Court in 1982.

During the dispute Rickard Deasy and others repeatedly exhorted farmers to avoid violence. In general this was respected. There was a clash with gardaí in Edenderry, where farmers attempted to 'black' sales to a Longford meat factory during the commodity strike (5 April 1967). There was a complaint that ten gallons of milk were spilt in Louth. As Fr Davis of Mayo said of the demonstrations, 'You would get more action in a country dance-hall of a Sunday night.'

Farmers' Response

The NFA declared a boycott of public events where any government minister was in attendance. The RDS asked that their Spring Show be exempted. The NFA responded that there was little humour for festivities among farmers whose comrades were in jail. They suggested that the RDS show solidarity with farmers by refusing the Department exhibit.

On the eve of the Show the farmers were still in jail. Dr Morris, of Cashel and Emly, a man long involved with rural organisations, was in touch with the Minister for Justice and the NFA to obtain release of the prisoners. He had no success on the first day of the Show. The *Irish Independent* had a front page headline, 'The Show Without Farmers', pictures of stock-judging rings with a handful of people. As an Exhibitor said, 'I have seen opening days when crowds of farmers were five or ten deep.' Next day 28 farmers were released and the Show was thronged. Rumour had it that a 'well-wisher', thought to be the BGA, had paid the fines.

Ploughing matches continued to be held. The arrangement was that all farmers would cease to take part if any member of the Government appeared.

A gathering of 2,000 women met outside Portlaoise Prison to hear an address from TJ Maher and to sing to their men inside.

Neil Blaney, introducing the Estimates for his department restated the Government position: 'It is regrettable that one farming organisation – I refer, of course to the NFA – should have elected to depart from the normal procedure of consultation with the Government and to embark instead on a challenge to orderly democratic government. The facts prove conclusively that responsibility for the present situation rests solely on the NFA.' Last August 'it had become perfectly clear that the only objective of these people was to put themselves in a position where they could, by one means or another, dictate to the Government. No Government worthy of the name could tolerate this kind of thing.' The Association's campaign did not represent the feelings of more than a handful of farmers – most of them very large farmers indeed, he said.

On the same day Rickard Deasy and TJ Maher were addressing the six hour NFA National Convention, with 3,500 attendance in the 2,200-seat National Stadium. Deasy said:

It is the Government and the Government alone that has at every turn attempted to present this struggle as a struggle to uphold the rule of law.

To this end they have sought to doctor radio and T.V., to fabricate fictitious incidents, to construct false rumours, to create new, so-called farm organisations, on a purely political basis, to foster and foment hatreds, hatreds for individuals, hatred for NFA. They have even stooped to twisting and misapplying the law itself in the name of justice.

To present our drive for a National Small Farm Development Plan, for a Meat Board, for a new animal breeding programme as a threat to democracy is nonsense – it is also a lie.

All we ask is that they leave the purely vocational role of vocational organisations alone, that in purely vocational matters they give us a just hearing followed by Government action, speedy enough to match the reasonable requirements of a rural and farming community that is in the midst of a technical, economic and social revolution of growing momentum ...

There does not seem to be the slightest use in attempting to communicate with a Minister for Agriculture who has devoted himself to the destruction, not only of the NFA, but to the permanent debasing of farmers generally to the role of yes-men to party politicians.

But to the Taoiseach we say that we recognise the need to reconcile the functions of the central authority and of vocational producer groups, that we believe that this is the root of the problem of democracy today.'[20]

At this meeting that Farm Business Development (FBD) was launched. The funds arising from insurance and other business ventures were to be used for agri-business development. It was a relief for practical men to undertake constructive organising instead of the negative dispute with Government. It was also seen as recognition that the dispute had become endemic – there was no apparent way out of the impasse. The farmers had better get on with their own business without government intervention. Neither side relished that prospect.

Next day Mrs Fields, of Kilcock, organised a women's rally. Sean Healy met the Taoiseach privately to smooth their path. Lynch was prepared to meet the women provided the rally was half a mile from the Government Buildings. The meeting, at the 'Peppercanister' church (400 metres from Government Buildings) was addressed and encouraged by Paddy Lenihan, father of the Minister for Justice. Six wives of imprisoned farmers (two of them in advanced pregnancy) had what Lynch described as the most constructive discussion yet on the dispute.

The following day the Government, after a Cabinet meeting, announced that all fines due from NFA members would be remitted. It was indicated that, if members still refused to sign recognances, they could appeal to the Circuit Court. The appeal was successful on 12 June, releasing 41 farmers. Bonfires blazed along the road of the cavalcade of cars on the journey to the Market Square of Navan. A crowd of 3,000, with TJ Maher, the newly elected NFA president, welcomed them.

Peace

Peace had come by a happy misunderstanding. Rickard Deasy issued a carefully crafted 'personal message' that members should pay their taxes if they could, but should clear all their debts to merchants and others first. Traders were partners in production. If traders were not paid they would lack working capital and be unable to extend the usual credit for purchase of farm supplies next year. If commercial debts were paid first the Government would be forced to the unpopular seizure of the goods of the poor.

The Taoiseach heard the 8 a.m. News while shaving (2 June 1967). He took the simple message that the rates strike was over.[21] Local elections were due on 28 June. Though the Fianna Fáil party denied any grass roots pressure, the ground swell that was to sweep the BGA and the Dublin Milk Board was already felt. To show that the Government wanted peace he arranged the release of the prisoners. The attitude of NFA Executive had so hardened by this time that Deasy could not have got agreement to conciliation but for Lynch's happy misunderstanding of the News.

Widening the Dispute

Neil Blaney, Minister for Agriculture, did not share the Taoiseach's views. He introduced the Livestock Marts Bill (19 June 1967), provoking long and stormy debates in Dáil and Senate. The NFA had long sought the aid of banks and government to limit the wasteful and inefficient spread of small marts. Neither the NFA nor the IAOS, organisers of most marts, nor the Association of Livestock Marts (which covered private marts) had been consulted by the Minister. The Act gave arbitrary powers to the Minister to grant or refuse licences; to grant exemptions; impose different conditions. In spite of farmers' representations President de Valera signed the bill. On appeal by the NFA, the Supreme Court found the Act infringed the constitutional guarantees of equality before the law and the separation of the powers of Executive and Judiciary.

1968

Apparent easing came when TJ Maher met Neil Blaney socially at dinner during the Cork Agricultural Show, shortly after taking office as President of NFA. He reported: 'As the Minister was shaking hands with me he was at that very moment removing NFA members from boards, semi-state bodies etc. and was withdrawing advertisements from the *Irish Farmers' Journal*. I wish to reiterate my statement, which I have made so often, that with an indication of sincerity from the other side, I am ready to go more than half way to meet them.'[22]

The Minister's displeasure spread to all groups considered favourable to the NFA. He would not meet the President of Macra; Department officials could not judge at Macra livestock competitions; all Department advertising was withdrawn from the *Irish Farmers' Journal* – an action making dangerous enemies of all journalists. In the Dáil he was referred to as 'Minister for War'.[23] *The Democrat*, a paper consistent supporting Fianna Fáil in the Minister's constituency, made editorial comment:

> The order banning an influential organ of public opinion from its share of Government disbursements is more than an indiscretion. It is an Injustice'.
>
> 'On what principle does Mr Blaney base his decision? He cannot quote a right. Mr Blaney, or any Minister, is not the owner of the public purse. The Government are not the masters of the people, they are their servants, entrusted with the obligation of administering the laws of the land fairly and impartially ... *The Farmers' Journal* has committed no crime. It has exercised a democratic right to criticise according to its obligations to a large section of the community ... we look to the intervention of the Taoiseach to make the order void and, if possible , repair the damage that has been done to his party's prestige.'

Action against Co-operatives

The recommendations of Dr Knapp in 1964 for consolidation and reorganisation of the dairy industry had been welcomed and accepted. Rationalisation and the take-over of the Dairy Disposal Company by the co-operatives was proceeding, guided by the IAOS. However, the IAOS Council had passed a unanimous resolution of support for NFA actions and had provided facilities for the NFA demonstration. The Minister commissioned a new report by two Americans, Cooke and Sprague, aided by a 'liaison group': James O'Keeffe, later president of ICMSA, represented the NAC, G. Deignan, manager of the small Letterkenny co-operative in the Minister's constituency, P. Carrick of the Dairy Disposal Company, C. MacCarthy of the Department, G. Tierney, economist with IAOS. ICMSA and NFA were at one in denouncing this 'unnecessary report'.

The new report reversed the proposals of Knapp. The Dairy Disposal Company should be expanded and given new responsibilities in reorganising the home market, because 'we see little disposition among the co-operatives to unite and do the job'.[24]

Subsequent development of the co-operatives, their take-over of the Company and consolidation into billion pound international enterprises did not bear out that judgement.

In May 1968 the Minister wrote to the Chairman of the Farm Apprenticeship Board saying that he would agree to meet the NFA President. Immediately a new setback arose. The Minister intervened in the election to the BGA on political rounds. Martin Corry TD, of Fianna Fáil, was a long-standing chairman. TJ Maher intervened, calling 'on all farmers to safeguard their right to a free election. The NFA candidates were elected and the meeting with the Minister did not take place.

Demise of the NAC

Attendance at the NAC was, perforce, political. The County Committees for Agriculture, appointed on a party political basis, agreed to attend. Intensive NFA canvassing of county councillors and some swing in local election voting persuaded them to withdraw in 1967.

The beet growers protested their strong objection to the snap decision of their BGA Council to appoint representatives to NAC. The NFA National Council supported their members objection and helped to contest the BGA election.[25] Fine Gael and Fianna Fáil made common cause against them. Gerard Sweetman, Fine Gael, warned Sean Healy that the BGA was political territory and that the NFA should keep off.[26] NFA sponsored candidates had almost complete success, drawing 90 per cent of the votes cast. Only two of the old Council were elected. Another ten votes would have given a full victory.

In the Dublin District Milk Board elections NFA candidates filled all seven seats.

The NAC now consisted of the government nominees and the ICMSA. The ICMSA co-operation withe the Minister was considered by the NFA as an act of treachery, weakening the farmers' struggle for independent representation. Like the passing of ICMSA pickets by NFA it embittered relations for years. The *Irish Dairy Farmer* described NAC as a weapon to drive farmers apart: 'its decisions are suspect, its writ without honour, its members without glory' (April 1967). At the meeting of 21 January 1969 the Taoiseach agreed that the NAC was dead and that it was up to the farmers to decide who should represent them.

Renewed NFA Farmers' Rights Campaign

The NFA launched the first phase of their renewed Farmers' Rights Campaign by placing pickets on the Department offices during office hours on 27 August. Apart from a few days they remained till November.

ICMSA was invited to joint action. A number of meetings, in the Clarence Hotel, proved abortive because the ICMSA insisted that their only mandate was to obtain an increased milk price while the NFA wanted a range of discussions. The two organisations 'unanimously agreed that no invitation from the Minister could be accepted until he indicates that he is prepared to consider favourably the urgent case for a milk price increase ...While other sections of the community are being compensated for rising costs the dairy farmer has been ignored.'[27]

The next invitation from the Minister to the NFA: 'was couched in such arrogant language, repeating Government allegations and issued to the press before being issued to NFA' that NFA could see no way to accept it without surrender.[28] The ICMSA also refused the invitation.

A thousand farmers, lead by TJ Maher and Paddy Lane, marched from Adelaide Road to to Government Buildings to picket the Cabinet meeting of 24 September.

When the Taoiseach stated his willingness to meet NFA (7 November), ICMSA requested a postponement because their President, John Feely, had died and could not be replaced till December.

TJ Maher's address to the 13th AGM (21 November) could only report failure for all attempts to co-operate with the Government during his first 15 months in office. The Taoiseach had launched the National Productivity Year 'with the involvement of all sections of the community' but the organised farmers were not even on the committee. The NAC had disappeared, leaving no governmentally recognised organisation representing all farmers.

Transition to Normality, 1969

The NFA resumed constructive work. By January 1969 £3 million capital, the stated target, had been collected to launch the FBD, enabling it to assist the take-over of Irish Meat Packers by Cork Marts. The appointment of Alan Dukes as chief economist, with Allan Matthews on the staff showed a change in priorities.

TJ Maher had a private meeting with the Taoiseach, Jack Lynch. The Government was now prepared to recognise NFA from the point of view of full consultation and negotiation. There was no question of the Government trying to impose a structure by way of a council on the farmers. The NAC was dead. The Minister for Agriculture had been told to put the *Irish Farmers' Journal* back on the list for Department advertising. The Executive, with reservations and not unanimously, agreed to meet the Minister.[29]

The establishment of the Livestock and Meat Promotion Body was announced (9 January). The Taoiseach asked NFA to meet the Minister to examine the draft text of the Third Programme for Economic Expansion (9 February). The Liquid Milk Group met the Minister in May. Restrictions on transport were greatly relieved when the Minister for Transport and Power, Brian Lenihan, removed carriage of cattle, sheep and pigs from the restraints of the Road Transport Acts.

The Small Farm Scheme never materialised. It was overtaken by the development of EEC agricultural and regional policies.

Haughey and Blaney were dismissed from the cabinet for reasons not associated with agriculture (6 May 1970). The new Minister, James Gibbons, was brother of Michael Gibbons, prominent in NFA and IAOS, whose property had been seized in Kilkenny.

CONCLUSION

Like most wars, this dispute sold never have happened. All parties and the Irish community were the losers. Hundreds of man/hours were wasted by farmers, with similar waste on the other side. Perhaps 200,000 farmers and family members were involved in demonstrations; 200 went to jail. Most serious was the five-year disruption, even cessation, of the work of preparation for the wider Common Market; for marketing boards; for reorganisation of creameries and marts.

It was a civil war at a moment of national crisis.

The Government established, what had never been questioned, its right to govern. The jockeying for position between cabinet ministers made it appear that the object of government was personal achievement, where Lemass had pursued objectives of policy.

Neil Blaney had consolidated the NFA, as James Dillon's proposed reduction in milk price had established the ICMSA in 1951. Austin Deasy, when Minister for Agriculture, summed it up: 'There are three most powerful organisations in Ireland, the Catholic Church, the Gaelic Athletic Association and the NFA. Only the NFA can demand to see the Taoiseach at will.' Most of the other improvements were due to the adoption of continental systems rather than innovation in Ireland.

The developments would have been more fruitful and profitable five years earlier.

11

Boom and Bust

ECONOMIC BACKGROUND, 1970–1978

NFA leaders had frequently warned that EEC offered no bonanza. None the less, the Common Agricultural Policy (CAP) *promised* the Irish farmer almost everything he could hope for: i.e. access to the home markets of the industrial countries; prices equal to the native farmers of industrial countries; no quotas or other distortions of the market for raw materials or for agri-business; regional development policy to bring farm investment and social services to continental standards; a say in the formation of policy at home and on the export markets.

Now they could stand on the inside, hoping for the fulfilment of the dream of an equal standard of living with urban workers. George Colley, Minister for Finance, said in the debate on entry to the European Communities: 'Family Farm Income will be double its 1970 level, and, with further price rises that are certain in the Community, the actual increase in income will be considerably greater'.[1]

THE BOOM

The honeymoon years, 1973 to 1978, provided the greatest boom of Irish agricultural history. Any upset to the market was out-weighed by the general rise in prices concealing, for the moment, the rocks ahead.

- Gross Agricultural Output rose 183 per cent in value and 31 per cent in volume. Agricultural Exports rose ten-fold in value.
- Markets diversified. Britain's share fell from almost 75 per cent to 47 per cent. Even exports to Britain were now mainly at prices controlled from Brussels.
- Farm investment doubled to £1,750 million in current prices between 1970 and 1980 (equal to £3,969,000 at 1993 prices).
- Land prices rose from £413 a Hectare in 1972 to £5,844 in 1979. Every farmer became solvent; every banker eager to lend.
- Family Income per person in agriculture rose 30 per cent in real terms. From 80 per cent of an urban industrial worker's income it rose to 10 per cent above in 1973 and peaked at 15 per cent above in 1978. (The comparison is rough, but shows the trend).

The most optimistic forecasts were fulfilled.

Michael Malone, Chairman of the National Milk Group, stated in 1971: 'It is not often, especially in agricultural circles, when one can say that we are completely satisfied with prices. This in fact was the position at least for a time this year. The increase of 14 pence in old money, was certainly phenomenal and something which we would only dream about a few short years ago.'[2] Milk price rose from 16 pence a gallon to 64 pence in 1980 and 92 pence in 1985.

'In 1972, for the first time since records began, income per farm worker in agriculture, as calculated by the Central Statistics Office, increased to almost the level as that of industrial workers, while in 1973 per capita family incomes were higher than those in industrial employment.'[3]

The Burst, 1978–1983

The tide turned. The farmers' 'Terms of Trade' (price of farm produce compared with costs) disimproved.

Table 9 Comparison of Agricultural & Other Prices, 1972–1978–1983

Index	*Output*	*Input*	*Consumer Price*	*Food Prices*	*Agr. Wage*	*Land Price*	*Interest. %*	*Bank Loans to Agriculture*
1978=100								
1972	32	61	63	60	58	-	7.5	£103 m
1978	100	100	100	100	100	100	13.2	468
1983	142	179	209	179	197	36	15	1,088

Compared with 1978, the peak, the farmers' position had deteriorated. It now took 22 per cent more output to buy a given quantity of farm inputs; 34 per cent more to buy consumer goods; 26 per cent more to buy food in a shop; farm workers' wages had become 29 per cent heavier. By contrast, land had fallen to 36 per cent of its former price and farm family income by 55 per cent. Farm borrowings from banks and ACC rose from £97 million in 1970 to £1,546 in 1983. Debts, incurred in the 1970s, when interest rates were lower than the rate of inflation and land prices were rising, became a serious burden when interest rates rose to 15 per cent and farm prices fell. It took more gallons of milk to pay the interest.

To some extent all this was the result of market forces outside Ireland's control. OPEC increased the oil price for every country. Agricultural prices were increased 28 per cent within EEC in partial compensation for the oil shock.

Ill-advised policies made the Irish situation worse than in the rest of Europe. The link between Sterling and Irish Pound became most damaging in an unforeseen way. Britain was forced to devalue by some 40 per cent. Ireland, with rising export prices, was under no such pressure. Prices rose steeply except in the case of agriculture. Farm prices were held down by monetary compensation payments (see chapter 14). Increased output and reduced numbers on the farm compensated to some extent for the adverse price changes.

Expectations that the food processing industry would grow strongly in a free European market were disappointed. Under CAP agricultural prices were set in national currencies. Devaluation did not affect this price. Food processed beyond the first, bulk, stage was 'industrial', so it changed with devaluation. Consequently, when currency was devalued farm produce, the raw material of this industry, became cheap but the finished food could rise. British food exports to EEC increased from £23.3 million in 1972 to £313.9 million in 1978. Ireland obtained a devaluation of the Green Pound. This gave relief to the farmer but almost exterminated the high value added food processing industry.

Irresponsible Financial Policy

The Government attempted to spend its way out of the world economic depression.

Uncontrolled demands brought wage increases of 80 per cent per unit of output in Ireland compared with 28 per cent in the more stable and better governed Continental countries. Irish government expenditure increased 142 per cent, compared with 42 per cent on the Continent. Increased taxes took away most of the advantage of increased wages and productivity in Ireland – but still was insufficient for Government's profligacy. Government debt became four times tax revenue and interest rates rose. Had the debt been incurred for productive investment all might have been well, but the money was largely used for public service wages and for social purposes, reducing the savings available for investment and future growth. The burden of government was a major cause of poverty and unemployment; the greatest attraction to industry was relief of taxation on exports.

This rather prolonged description of the economic background is necessary for the understanding of the changed and newly acrimonious relations between farmers and Government in the 1970s and 1980s.[4]

Though the highest hopes of development in the Common Market were disappointed, it does not follow that Ireland's entry was a mistake. The alternative 'World Prices' obtained by New Zealand and other outsiders were a fraction of those in Europe's sheltered market. Free trade in manufactured goods brought industrial expansion impossible for a small isolated state. There was a clear net gain.

IFA Reaction

Government and farmers were broadly agreed on foreign policy. The Irish feared to take any initiative in reform of the Common Agricultural Policy lest any change would be for the worse. The IFA recommended that the increase in prices to continental levels should be spread over a longer transition period to reduce the shock to the market. The Government chose the immediate price increase and the postponement of quotas for as long as possible to allow increased output to be established. Efforts by farmers and Government to abolish MCAs were unsuccessful. There was no use talking economics when the UK and major EEC partners had made a political arrangement. The IFA National Council discussed breaking the Sterling Link a number of times but, as Lorcan Blake, IFA economist, pointed out (June 1976) this could not work when Irish inflation was even higher than the British.

The relative position of farming was worse than in 1972.

25 Taxation drew off the capital needed for expansion.

12

From NFA to IFA

The renaming of NFA as the Irish Farmers' Association (IFA) in 1971 had more than a symbolic significance. The vision of John O'Dwyer, chairman of ICMSA, that there should be one organisation representing farmers, faded at the Thurles meetings of 1953 and 1954. None the less Juan Greene's sketch of a truly representative body, which he had put before the Thurles meeting in 1954, was coming close to reality by 1963. Fifteen commodity associations came under one umbrella. They retained their identity but worked as committees with representation on the IFA National Council and eventually, in 1973, sharing offices in the Farm Centre. Their policies had to be endorsed by the National Council. The road to unity was not smooth.

While a very large number of local leaders emerged in the national movement, the same names occur in the different organisations at each level. Certain families have a tradition of farm leadership for generations. Table 10 illustrates a few individual cases.

Table 10 Some Examples of Interlocking National Leadership of Macra, Co-ops and Union

Name	*Macra*	*NFA/IFA*	*ICOS*	*Other*
J.N. Greene	President of 1st Club	President		V. President IFAP
Pa Quinlan	President	Inaugural Chairman	President	
J. Litton	Chairman			IFJ Board: Teagasc
J. Mooney	President			Chair IFJ Board and FBD
H. Meredith	Chairman			Chair LMP
P.I. Meagher	President		President	Chair IAWS
M. Gibbons	Chairman		President	Chair NFA Linstock Ctee etc.
W. Carroll	V. President		President	
J.F. Orpen		V. President		Chair Grain board
A. Gillis	President	President		Chair LMP: MEP
J. Rea	President	President		Chair Teagasc
TJ Maher		President	President	President CAPFAR: MEP

Beet Growers' Association

The BGA position as representative of the 30,000 beet producers with the Sugar Company was enshrined in the Act of 1926, together with its legal right to collect a levy on the beet crop. A Liaison Committee was formed with NFA in November 1955. BGA worked so closely with NFA in grain that the BGA chairman had a seat on the NFA National Council.

The political associations of BGA presented problems. The crunch came during the Farmers' Rights campaign when the politically elected Council sent representatives to Blaney's NAC without consulting farmers. Beet growers in Carlow (led by Jim Kehoe and Jim Mullins) appealed to the NFA Council,[1] for help in fighting the BGA Council election. The result was the dismissal of the old Council members (see chapter 11).

The BGA, now the Beet and Vegetable Growers' Association, retained its separate identity but acted as a committee of the new IFA. The BVA officially combined with IFA in July 1979, though the integration was rather loose. Contrary to the intention at the time, BGA retained its independent finance. As a result the expenses paid to BGA members for attendance at meetings was 75 per cent higher than those paid to ordinary IFA members. When IFA could not find money to send one representative to the IFAP World Congress in Mexico the BVA could send its president, secretary and three vice-presidents to a sugar conference in USA. Full integration was announced in 1991. In case of disagreement, the Beet Committee could retain its money. The 1996 AGM extended the IFA election system to the Beet Section, marking BVA's final integration into IFA.

Liquid Milk Producers

Producers of milk for direct human consumption in Dublin and Cork were a special case in farming. In recognition of their licensed premises and the costs of year-round production the law gave farmers in the 'designated areas' exclusive rights to supply the major cities. These farmers had much in common with city workers. The Leinster Milk Producers' Association, formed in 1934, could enforce a closed shop. As suppliers of an essential food liquid milk producers could put immediate pressure on consumers by withholding supplies. They bore the brunt of commodity strikes, making common cause with the ICMSA in 1953. It would have been more natural that they, rather than the ICMSA, would stand aloof from the national association.

Juan Greene called a meeting of all liquid milk associations in 27 Earlsfort Terrace on 14 April 1960 to discuss the deteriorating prices proposed by the Government. He argued that the overall national interest would best be served if they approached the Government through a co-ordinating committee – while keeping their individual organisations intact.[2]

At the second meeting of this liaison committee a letter was read from Patrick

Ryan, secretary of LMP, regretting the absence of the ICMSA from the committee. The LMP, he wrote, was conducting discussions with the ICMSA, with whom they had more interests in common. Since the Dublin and Cork prices were government controlled, 'the problems of the various outside organisations are not on all fours with ours'. The older leaders of the Leinster Milk Producers were loyal to their allies of past strikes. They feared that they might be swallowed up by the general organisation and their interests ignored.

The Cork Milk Producers' representative expressed his disagreement with these views. His group, as always, favoured working through the NFA.

However, influential farmers were members of both NFA and LMP. Harry Meredith of Wicklow, for example, who, as chairman of Macra, was principle speaker at the foundation meeting of NFA, was also a former chairman of LMP.

In April 1962 the NFA again met the LMP. In the intervening years there had been a power struggle within LMP, which now expressed the unanimous desire of its Executive for close working relations with the NFA. Alfred Melvin, chairman, took a seat on the NFA Council and two LMP representatives joined the National Liquid Milk Group of NFA. Fees for both organisations were collected through the wholesalers and the same canvassers called from house to house to enrol members. Questions of milk standards and quality payments were discussed and the foundation of a National Dairy Council to promote the conscription of milk by advertising. The NFA economist (Louis Smith) advised a co-operative structure for Dublin's pasteurising and distribution.

LMP remained independent. When the NFA proposed a National Milk Board to control contracts, quality standards, prices, and sale of surplus milk in 1966, the LMP disagreed. It would prefer the existing Dublin District Milk Board to such a watered down organisation. It wanted a more powerful, producer controlled board. Again, during the Farmer's' Rights campaign the executive of LMP pointed out to the NFA that no outside body had a right to call a strike in the Leinster area, and no such mandate had been given.

During the NFA Farmers' Rights Campaign farmers from all organisations were imprisoned together because of the machinery demonstrations. The common experience and daily contact cemented friendships. The Organisation in Mountjoy, described in Chapter 5, was duplicated by similar groups in the prisons of Limerick and Portlaoise.

LMP's connection with NFA became ever closer. It subscribed £525 towards the building of the Irish Farm Centre, as a base for all the farm organisations in 1969; joining in the formation of IFA in 1971.

Organisations of suppliers to the smaller cities Limerick, Waterford, Wexford, Dundalk, The Midlands merged in the NFA Liquid Milk Group.

Women in Farming

The Irish Country Women's Association continued its long established activities in friendly independence. Integration was not contemplated. (TJ Maher's application for membership of ICA was refused.) Women are active in the business of farming as owners or, often, as the business head of the family. It caused headlines when a woman was elected to the first Dublin District Milk Board. By 1955 NFA membership was by families; wives and daughters were automatically members. Any member of the family could attend meetings. A resolution by the National Council: 'That the NFA welcomes the naming of 1975 as International Women's Year and gives its support to the programme to draw attention to the role of rural women in our society. In particular, the Association calls on farm women to actively participate at all levels in this Organisation.' Nonetheless, the chairman of the Sligo district artificial insemination service adjourned the meeting immediately Sheila Hamilton appeared as a delegate. When approached to be president of ICOS she felt society was not yet ripe for a woman's candidacy.

The contribution of women was particularly evident during the demonstrations of 1966, when 2, 000-strong sang outside the prisons. Their march through Dublin and interview with the Taoiseach, Jack Lynch, was crucial in terminating the NFA dispute with the Government (see chapter 5).

Women were more clear-sighted than the men in political matters. Asked why they were confident that Alan Dukes, Minister for Agriculture, would, fulfil his promises, they replied that they had confidence because he knew they controlled enough votes in Kildare to ensure his defeat in the next election. Had the farmers of Waterford and Kerry had the same vision in 1966 they would have saved themselves, and all Irish farmers, a deal of anguish.

In November/December 1975 the Farm Family Group was formed by the National Council. The committee was particularly concerned with questions of water supplies, education, health, farm safety and social problems. This committee represented Ireland in the Feminine Committee of COPA.

The most significant development is that women are now elected to the National Council in their own right as farmers, not as a special appointment by concession.

IAOS/ICOS

In Sir Horace Plunkett's programme the co-operatives were to represent the farmers in all sorts of ways – business, education, negotiations with the state and other bodies. In NFA thinking there was a clear distinction between co-operatives, charged with the ownership and management of business, and a union engaged in negotiation of policy with Government. Co-operatives are vulnerable to government action. In the Irish case the IAOS was too dependent on government subsidies to lead a forthright dispute. The IFA favoured the division found in Sweden where the two arms of farm organisation are complementary but separate.

Even in Macra the future leaders of the NFA had been active in forming co-operatives for wool marketing, purchase of fertilisers and other activities. The IAOS serviced the co-operatives but was relatively weak in grass-roots contact. Here the network of 500 Macra and 900 NFA branches spread knowledge of co-operative projects and canvassed door to door, raising capital for marts, FBD or other projects.

Designed in the days of horse drawn transport, the co-operative creamery structure, creamery reorganisation was overdue. In the 1950s it was common knowledge that the costs of manufacture were 30 per cent less in big creameries than in small,[3] and the reorganisation in Denmark, Netherlands, Sweden was studied.[4] Government policy did not favour consolidation. When the IAOS, with great difficulty, organised the amalgamation of Ballyclough and Kilcorney creameries in 1971 the Department forced a 'divorce', apparently for sociological reasons. Re-equipment was necessary, 50 per cent of equipment was over 50 years old, but it was wasteful to give grants for equipment in redundant creameries. Firm NFA support was given to the IAOS 'Framework for Co-operative Development' (1979) in rationalising the creamery structure.

Development of large co-operative business controlled by farmer directors, without business training or experience, seemed foolhardy. To invest in such ventures or to make large unsecured loans to them would tempt Providence. That was the message clearly spelled out by the Agricultural Credit Corporation,[5] the Cooke and Sprague Report on the dairy industry and many times by merchants. The newly amalgamated creameries brought together men who were strangers to each other. Fellowship and local solidarity, the strength of the co-operative system, had to be built. To make matters worse, most local creamery committeemen wanted to be on the board of the new regional co-operative. Executive boards of 80 resulted. Such a board could not make dynamic decisions.

At the IAOS AGM Louis Smith, consultant economist to Allied Irish Banks, conveyed an offer from the bank to fund short courses for co-operative directors. These were held at weekends away from home. Free from daily chores the farmers and managers could get to know each other in a new way. Lectures and discussions on business methods, work of boards of directors were conducted by co-operative managers and by academics from business schools. Co-operative principles and the objectives of their own societies were developed. At the commencement committee views were different even on the basic objectives of their organisations.

These were not business meetings. the results cannot be measured. The establishment of a core group of a few thousand directors, exposing them to new management ideas and making friendly contacts between them, contributed to the rapid, successful growth of recent years. There have been no bankruptcies.

The co-operatives were invaluable to the NFA in collecting membership fees among their 100,000 shareholders through deductions from the milk cheque or by a levy on cattle market sales. The comfort given at Plunkett House to the deputation waiting outside the Department in October 1966 was appreciated.

NFA officers came to the fore in ICOS because of NFA activity at the grassroots in promoting new co-operatives and raising shares for the development of those already established. In any election the publicity given to NFA leaders attracted votes. NFA interest was also stimulated by the need to defend itself. The NFA minutes make frequent reference to the efforts of ICMSA to prevent collection of NFA fees through creameries. To defend itself and to promote co-operation the IFA put forward candidates for election in local co-operatives and in ICOS.

ICOS did not move its head office to the Farm Centre. It continued in its prestigious head office, 84 Merrion Square, presented by subscription of well-wishers when the Department cancelled its grant in 1907. There was no urgent need to move to the suburbs. In the 1970s some IFA leaders took an ambivalent line on co-operatives

As in the Scandinavia model, there was never question of combining co-operative business with the farm pressure group.

The co-operatives responded to the new opportunities with phenomenal growth from £20m turnover in 1970 to £6,000 million in 1995. Employees now number some 25,000. They dominate all sectors of agribusiness except the dark area of the meat trade.

Other Associations

The Agricultural Association of Ireland, formed before the NFA, worked closely with it. They shared office space in 27 Earlsfort Terrace, the AAI continuing to negotiate agricultural wages, using the research of NFA in an amicable relationship from June 1955.

At the 1955 AGM of the Irish Farmers' Federation it was unanimously decided to dissolve the Organisation as from 29 June, in view of the effort now being made to organise farmers under one national association. They wished the organisers complete success. IFF had held the fort for 17 years and hoped the younger generation would not forget.

Smaller associations came under the IFA umbrella. The Bee Keepers' Association, with a thousand members; the livestock breed associations; the Irish Horse Breeders; Irish Commercial Fruit Growers ...

Other groups had good reason to remain independent but in close association. Macra na Feirme and its kindred junior movements, Macra na Tuaithe and For Oige, were part of the same movement. The President of Macra became, ex Officio, a member of the NFA National Council. Similarly, Macra placed a representative on the ICOS Council. The ICMSA, having no conflict of jurisdiction, also placed a member on the ICOS Council.

Ulster Farmer's Union

A special relationship between Macra and the YFCU(Young Farmers' Clubs of Ulster) matured when the NFA was launched. Formed in 1918 the UFU had 30,000 members in 250 branches, with paid secretaries for groups of branches. The members and branches were kept in touch by the *Farmers' Journal* (of Ulster). The NFA at first presumed that they would enjoy similar harmonious relations of UFU (and the English NFU) with Government and marketing boards. The experience of the older organisation was made freely available. Their officers spent time attending meetings in the preparatory stages of organisation. At the launch of NFA in January 1955 Arthur Algeo, president of UFU, and Humphry Jameson, general secretary, were platform speakers. Many UFU members made themselves available at local meetings when branches were being formed.[6]

In September 1961 the institution of a Joint NFA/UFU Consultative Committee was announced. The Committee met every three months, venue and chairmanship alternating between the two organisations.

Again, when the NFA were deeply concerned at developments with the Government in 1966, John Martin, president of UFU, addressed the Special Council Meeting (8 September 1966) urging farmers to control their own business. Closeness of association rested on the common interest. The land border enforced a common livestock and disease control policy; Irish bacon was always known in the market as Irish bacon; the small farms of Ulster, North or South, had more in common with each other than with the large land owners of England when discussing the Common Agricultural Policy in IFAP in Paris and in COPA. Underlying it all were strong personal friendships between Johnny Martin, Harry West, Sean Healy and others, developed as home guests when on the regular meetings. Further co-operation naturally followed.

Relations with the ICMSA

The tangled relations between the IFA and the ICMSA now extend over more than 40 years. There is little good to be said about the matter. No two people involved would tell the story in the same way, but some analysis of their differences is necessary because their rivalry recurs at almost every stage in this history.

There is a difference in origin colouring their approach to every question:

The ICMSA began as a pressure group of creamery suppliers, founded in 1949 because of discontent with dairy farmers' incomes in Limerick (see chapter 1). James Dillon's proposal to cut milk price by one old penny a gallon (8 per cent),[7] seen as a threat to every dairy farmer's income, made the ICMSA a national force. Political pressure had to be their chosen weapon because the price of butter, and so the price of milk, were completely under government control. The price of milk was the dairy farmers' wage. James Dillon refused to have discussions with the new organisation (ICMSA): 'I state categorically now that I regard them as a tendentious political racket, that I do not intend to discuss with that association any matters relating to the creamery industry or anything else.'[8] He attacked them with vigour, while encouraging them with an increase of 0.8 pence a gallon.

A 16-day-milk strike in February 1953 was a trial of strength. The ICMSA advertisements were forthright in denunciation: 'The vote [in Estimates of Government Expenditure] for the Department of Agriculture is a vote for exploitation of agriculture. ABOLISH THE DEPARTMENT. LET THE FARMER RUN HIS OWN BUSINESS'. They demanded an inquiry into the cost of production of milk and recognition of the farmers' right to consultation.[9] The price increase of 2 pence a gallon was a clear victory, measurable in the monthly creamery cheque.

The ICMSA started with successful agitation. There was no preliminary development of policy, even in dairying. No full-time professional staff were hired till Jerome O'Shea joined in 1967. Dr Raymond Crotty was economic adviser for various projects from 1961, when he prepared the report on the prospects for Irish agriculture within EEC (see chapter 8) and on the 'two-tier' milk price. By contrast the NFA grew from the educational work of Macra. It prepared its policy by a year's intensive preliminary study in commodity committees of farmers, with academic and agri-business help. An economist was employed six months before the public launch. The first major policy intervention was on the proposal for the Agricultural Research Institute. It had, in the first years, no thought of strikes or demonstrations against the Government, presuming that close friendly relations found in other countries would develop in Ireland.

Interorganisational conflict between associations was not inevitable. Macra and the IAOS co-operated in promoting the foundation and recruitment of the ICMSA. The ICMSA responded positively to Macra's call for all farm organisations to form an overall farming Organisation in Thurles in October 1953. The ICMSA participated fully, sharing the chairmanship in rotation.

Two basic differences in policy emerged.

1) The BGA refused to exclude professional party politicians from office in the proposed association.

2) The ICMSA would not accept any overall body with power to co-ordinate policy or to speak for dairy farmers.

The ICMSA difficulty appeared to be overcome when John O'Dwyer wrote to the *Irish Farmers' Journal* that 'the only outstanding obstacle to unity' was the BGA's position on members of the Oireachtas.[10]

The ICMSA delegates were not empowered to commit their association to any agreement. They had to refer back to the Administrative Committee for ratification. Endless exasperation was caused when agreement was reached at a meeting, only to be set back by rejection from the Administrative Committee and a new set of proposals drafted. The rules of ICMSA gave exceptional powers to the small Administrative Committee of officers. Charles Fletcher, of Killeshandra as chairman, John Feely, hon. secretary, John O'Dwyer, hon. treasurer, were on the Administrative Committee. Liam Barry, Kilmallock was secretary. In August 1952 Fletcher became president and Feely chairman. When O'Dwyer was ousted, Feely combined his post as secretary with that of chairman. In 1961 Fletcher resigned and retired to the Isle of Man. The rules have an unusual provision that the Administrative Committee 'shall have power to veto any decision by the National Council, such veto to last for 12 months.'[11] The Administrative Council prepares the agenda of meetings and can 'if it deems a resolution irrelevant or not in accord with the objects of the Association, 'exclude such items from the Agenda.'[12] This concentrated power was to be curtailed by the rule that, 'a person having served two consecutive three year terms shall not be eligible for re-election until at least one year has elapsed.' The rules were drafted by Conor-John O'Callaghan-Westropp, of Clare, who was afterwards chairman of the NFA Rates Committee.

Unity or Federation? 1955

Agriculture was based on small family farms with mixed products. The ICMSA title inhibited its work outside the creamery areas and in lines other than milk. Conflict of interest between single product organisations, milk, store cattle feeders, grain growers, would be inevitable. For this reason O'Dwyer stressed need for an overall association and took the initiative in meetings to establish such an Organisation. The breaking point in Thurles was failure to decide whether the new Organisation should be a loose federation or a united body.

When Macra invited the ICMSA to the inaugural meeting of the NFA, a unified organisation, in January 1955, the reply was a counter-invitation. ICMSA asked for 19 Macra delegates to meet them under an independent chairman, Bishop Lucey of Cork. This invitation was received in February, when the NFA was already a month old. Perhaps it was prompted by the rapid take-off of NFA. Macra replied that it had reverted to its original social and educational role and passed the invitation to NFA.

Juan Greene stated his position (12 February 1954):

> Existing farm organisations have nothing to fear from the NFA, whether they decide to actively co-operate or not. They naturally will not allow themselves to be integrated within NFA without having an assurance that its members' interests will be fully catered for, and nothing is more reasonable. Yet 95 per cent of all farmers believe in the ideal of a national, comprehensive organisation, so let us contribute individually to its achievement and so earn for the NFA the confidence which must one day lead to the integration of all Farm

Organisations within its framework. A little time may be necessary to reach this stage, but again I believe reason dictates that it must happen.

The ICMSA, after due consideration, have acted in what they believe is the best interest of their members. We, the NFA, may regret this decision, but we must accept it. Members of the ICMSA will, I hope, stand behind their Organisation, which has served them so well, and if, according to the dictates of reason, he can see in an NFA a further development towards even greater service on a much broader field his support and membership will be well worth while.

Apparently this statement was taken by John Feely as an offer of a place on the the National Council of NFA. He replied in a letter to the *Irish Farmers' Journal* as deputy president and hon. secretary of ICMSA: 'As regards the statement that I was invited to act on the first Council of the NFA, I want to make it it clear that, had such a proposal been made to me I would have refused'.[13]

Dr Greene elaborated his belief that the majority of milk producers required either 'a single comprehensive Organisation that would cater for all his needs and interests, or a commodity milk organisation plus another Organisation that would cater for his other interests. The NFA, naturally enough, recommends the comprehensive organisation as the best solution; but the milk organisations say, perhaps with good reason, that it is not a practical proposition yet. In deference to this view the NFA is willing to compromise and reach a solution on the basis of the second reasonable solution of having two organisations to cater for his needs. The NFA recommends that this can be achieved on the basis of one subscription for each farmer which would cover membership of both his milk commodity Organisation and the NFA.'

The ICMSA attitude developed. In 1954 the Tipperary chairman, Richard Ryan, said: 'The ICMSA will do nothing to hamper the NFA, but pending an agreement between the two bodies, members have been advised not to join NFA.'[14]

John Feely elaborated this view: 'ICMSA was the first to make a progressive move to bring all farming groups together on the basis of federation. Our efforts to achieve unity on this basis were thwarted by Macra na Feirme – the sponsors of NFA. In turn the NFA in a most arrogant manner attempted to gobble up the one and only body capable of protecting the dairy farmer.'[15]

The NFA and ICMSA met under Dr Lucey. There was no shift in the positions adopted. ICMSA wanted a loose federation of independent commodity organisations. NFA wanted united local branches, eventually uniting in one Governing Body. This would provide technical services and co-ordinate policy, enabling farmers to present a united front.

Compromise could not be found.

A final meeting in the Royal Hotel, Tipperary, 24 September 1955, made proposals to the National Executives of NFA and ICMSA:

1) Each organisations will retain its separate identity.

2) All members of the NFA who supply milk to Creameries will automatically become members of ICMSA.

3) All members of ICMSA become automatic members of NFA.

4) Only one fee to be collected from each farmer, which will be apportioned between the organisations in a manner to be agreed.

5) Both organisations will work together in organising in all areas.

6) ICMSA will look after the marketing, etc. of milk supplied for manufacturing purposes.

7) A joint committee will be set up for organisational purposes.

The proposals were to be confidential, not discussed at branch level, and a further meeting was to take place when views of the national councils had been obtained.

This was the nearest approach to agreement till 1964.

The NFA National Executive accepted in principle. The ICMSA discussed the proposals widely but never expressed an opinion on them. After a number of requests by NFA a further meeting took place in December (Clarence Hotel, Dublin, 19 December 1955) at which the ICMSA produced a new eleven-point proposal:

- The organisations should remain separate, with a loose federal structure at national level.
- There should be only one Organisation in each parish.
- Where either Organisation already existed the other should not form a branch.
- Where there was already duplication the newer branch should agree to close.
- The ICMSA subscription should be one shilling per cow and that of the NFA on a commodity basis.
- Only milk producers could become members of ICMSA.

The NFA, which had no sight of the new proposals before the meeting, could hardly have agreed. The proposal would effectively debar the NFA from all the prosperous dairying areas, where the ICMSA had already branches. Non- dairy farmers living in creamery areas could neither join the ICMSA nor form an NFA branch. The NFA would be required to service all commodities except milk and all other questions, such as fertiliser prices. Farmers in creamery areas, having no branch of the NFA to join, would be 'free-loaders', accepting the general benefits obtained by the the NFA while paying nothing.

John Feely clarified the situation at the meeting when he stated that the NFA should keep out of the dairying counties of Cork, Limerick, Kerry, Tipperary, Waterford and Kilkenny. Under the terms of his proposed agreement, NFA would also be excluded from all other creamery areas where ICMSA branches had been formed. It is unlikely that the proposals were expected to be a base for discussion. No public records are available on the background to the ICMSA approach.

In 1959, at a meeting in Croom, Juan Greene proposed to break the deadlock by the dissolution of the NFA Council and a general election to a new Organisation. Dairy farmers, being the most numerous section, would be well represented.[16] The offer 'which cannot be described as other than brave and well-intentioned'[17] was widely reported with favourable comment in at least 18 newspapers.

John Feely said he would have to confer with the ICMSA Executive. NFA has no record of any response. It was thought within the NFA that the chief influence was a fear that, with extended membership, officers of the ICMSA would be subject to change at elections.

Relations between the two organisations deteriorated from 1959 on. Few minutes of the Executive were without reference to the ICMSA. Reports from the branches told of, 'vigorous efforts against the NFA organisers in Monaghan' (January 1959)'; the ICMSA distributed printed forms for farmers to cancel payment of NFA membership fees through creamery or bank; the ICMSA publicity blamed NFA for the troubles of the wheat harvest (November 1962); defamatory articles in ICMSA's *Gazette* and other propaganda became vulgar abuse, considered 'beneath contempt' and not worthy of reply.[18]

Contacts in 1964

On taking up office in 1964 Rickard Deasy renewed relations with the ICMSA. Many meetings took place. Proposals were drafted, agreed by the NFA Council; rejected by the ICMSA.

An informal meeting[19] recommended to the two associations:

An Overall Liaison Committee with President and two others from each Organisation.

Functions:

a) to work towards harmonious relations between the organisations;

b) to eliminate all possibility of policy clashes or organisational frictions.

NFA to give ICMSA three or four representatives on NFA National Council two representatives on NFA Central Executive.

ICMSA to give NFA three or four representatives on its National Council two representatives on ICMSA Administrative Committee.

ICMSA to set up a National Dairy Committee with sole responsibility for originating policy on all matters affecting creamery milk and dairy products and should give NFA suitable representation.

NFA should afford ICMSA suitable representation on its Pigs Committee, at that time the only acting commodity committee of ICMSA.

Representation should be either:

a) equal on both committees; or

b) ICMSA 60 per cent of Dairy Committee and NFA 60 per cent of Pigs Committee.

ICMSA should also appoint two members to NFA Livestock and Rates Committees, other representation as may be agreed later.

It was essential that policy on all committees to be agreed by NFA National Council, which would include members of ICMSA.

A letter of Rickard Deasy to John Feely (19 June 1964) discussed the financial arrangements. Money from the creameries should be lodged in a joint bank account. The ICMSA should receive an amount equal to its 1962 income of £11,000 per annum. Any special expenses in the Dairy Committee should also be paid from this fund, up to say £15,000. The balance would go to the NFA.

The NFA should make all services, including economic advice, available to the ICMSA and fund all other committees. As of end July no answer was forthcoming.

In addition, the officers of the NFA repeatedly renewed Juan Greene's offer to retire en bloc, allowing the farmers free choice of leaders in a federated association. This was the arrangement followed with the associations federating with the NFA to form the IFA in 1974. The IFA suggestion of a separate voters list, allowing only creamery suppliers to elect the Dairy Committee (as the Beet Growers elected their committee) was rejected.

The key clause in all agreements was the intention to form a unified Association. To the NFA that meant a unified body able to reach decisions by majority vote. To the ICMSA it meant a loose federation, where decisions by the central council were not binding on commodity associations, which would remain fully autonomous in finance and in policy. The division was similar to that between Britain and the other members in the European Union.

At branch and county level there was jockeying for position. The ICMSA complained of intimidation in county Louth. The allegation was that milk cans belonging to farmers refusing to subscribe to the NFA were spilt by NFA supporters.[20] From the date of the incident it seems more likely that the dispute involved a farmer's supply of milk during the commodity strike. Any violent local action was contrary to NFA policy. Much of the time of Paddy FitzGerald, Limerick representative on the National Council and of Michael Howard, of Clare, was taken in counteracting ICMSA efforts to prevent collection of NFA fees through creameries and in forestalling ICMSA action by rival meetings of protest on rates or Shannon Development.

The harm done to the NFA by disunion and verbal abuse was not confined to Ireland. In February 1967 the Australian Primary Producers' Association, with whom friendly relations existed through Dr Greene's vice-presidency of IFAP, wrote to NFA. They reported that the former Australian ambassador to Ireland had made newspaper statements that NFA, 'was not on speaking terms with any other farm organisation at home or abroad'. This, and other statements, were considered not worthy of public reply, but answered by letter between staff.

Friendly personal links continued. For example Patrick Walsh, chairman of Cork NFA County Executive and member of National Council, was also vice-chairman of the Area Committee of ICMSA. Dr Crotty, economic adviser to the ICMSA, was influential with his brother-in-law Michael Gibbons, chairman of the NFA Livestock Committee. These contacts could neither soften the basic division, nor reveal that division's real cause.

1966

Deterioration of farm incomes in 1966 was felt equally in ICMSA and NFA. The ICMSA was further angered by the Government's continued failure to produce the findings of the Milk Costings Commission.

The ICMSA placed pickets on Government Buildings demanding an increased milk price. The NFA had an appointment with the Taoiseach to ask for the implementation of the Government's promise to adjust farm prices if the general rate of inflation exceeded three per cent. The NFA invited the ICMSA to come with them to raise the milk price question. The ICMSA refused any discussion unless an increased milk price had first been settled. The Depart ment refused to change the venue. The NFA chose to pass the pickets. In doing so they permanently soured relations with the ICMSA – and came away emptyhanded.

John Feely claimed that the institution of the Two-Tier Milk Price was the first instalment of the ICMSA Small Farm Plan, attacking 'the blatant hypocrisy of the NFA leaders who attempt to claim credit for the new scheme which they had done so much to hinder for their own selfish motives' (March 1968). The NFA Rural Development Section felt they had to reply. They pointed out that the ICMSA submission was solely on milk, with only the oblique reference to small farms, 'State aids to agriculture should be deliberately weighted in favour of the small farm sector.' The ICMSA had no plan for the generality of small farms.

Inevitably politicians reduced the pressure of the farm lobby by playing one association against the other. Charles Haughey, as Minister, was determined at first not to negotiate with farmers under pressure. Eventually he conceded a price increase of 8 pence a gallon, wording the announcement so as to give the credit to ICMSA. Later, Neil Blaney agreed to a 'two-tier' price for milk when the ICMSA took part in the NAC. The NFA was manoeuvred into passing the ICMSA picket.

The ICMSA was more successful than the NFA in its dealings with the Government during this period. This point was stressed by the Blood document (see chapter 5). In 1966 the National Council felt compelled to initiate 'negotiation by demo' because the members could see returns from ICMSA's agitation but none from the reasoned arguments of the NFA. This was a step backwards. It diverted the efforts of committees, NFA staff and Organisation to agitation at the expense of planning and construction. It was noted at the National Executive that the Economic Section had been sacrificed, reduced to one person, while field organisers were recruited.

1968

In August 1968 both organisations were picketing government offices. Farmers from the rival organisations met with friendly hand shakes. Press reporters asked the two Presidents for comment. Said TJ Maher: 'I welcome this. We have a lot of common ground. We're all farmers and our organisations are endeavouring to do the best we can for our members. Why shouldn't we shake hands? I would do it myself if I were there.' Said John Feely: 'I would not place any importance on it at all. There can be friendship between people but when it comes to business it can be different. There might be some hypocrisy in this. When they [the NFA] get out of milk there'll be no more problems.'[21]

In 1968 the ICMSA requested a meeting with NFA, the first formal contact since 1964. The death of John Feely prevented this meeting. However, TJ Maher reported to his president's advisory committee (26 November 1968) that he met the ICMSA representatives in Thurles. The NFA proposal that creamery suppliers should elect a new Organisation had been rejected.

Diplomacy and democratic elections brought the BGA, the various liquid milk suppliers and other commodity organisations into IFA, leaving only the ICMSA as a significant outsider.

Closeness, 1973 to 1980

TJ Maher saw in EEC membership an opportunity to bring about farmers' unity. He reported to the National Council on his discussions with James O'Keeffe, president of ICMSA (24 August 1972). On his proposal, IFA invited ICMSA representation on the European Affairs Committee and also on the four commodity committees then covered by ICMSA. He specified that the committees be subject to IFA Council and under the IFA membership of COPA. He had proposed that the agreement include a commitment to unity of the two organisations by a predetermined date, but the ICMSA had not agreed. The ICMSA wanted a loose co-ordinating body with rotating chairmanship. This was acceptable neither to IFA nor to COPA, the international body.

TJ Maher's presidential address expressed 'appreciation of the unremitting efforts made by the president of ICMSA (O'Keeffe) which led to the conclusion of this agreement' (26 June 1973).

The agreement worked satisfactorily (August 1974). 'I am pleased to be able to report that co-operation between the ICMSA and the IFA, through the operation of the European Affairs Group has grown and developed over the past year and a half in a most satisfactory manner. 'I want to acknowledge the contribution made by the president of ICMSA, Mr O'Keeffe, to this work. We are now engaged in discussions to map out the road ahead for the future, drawing more closely together than we have at any time previously. The importance of this argument, particularly in the context of European involvement needs no stressing from me. I believe all farmers recognise it.'[22]

In 1977 James O'Keeffe complained of the action of the local Limerick IFA during the Bord Bainne elections. The National Council agreed that electioneering was inappropriate in view of the relations between the two associations, who had agreed candidates. The NFA sent an apology.

Good relations continued under the presidency of Paddy Lane, 'over the past two and a half years we have seen a steady growth of co-operation between IFA and ICMSA in the European Affairs Group which was of great benefit to Irish farmers ... in the years ahead I hope to see this co-operation develop to a stage where both organisations come together to form a united farmers' association, dealing with all agricultural matters both in Europe and at home.[23]

None the less there was a continuing undercurrent of unrest. The ICMSA objected to the IFA's collection of funds for the European Affairs Committee through bacon factories, as agreed with the Bacon Curers' Association. The curers and the IFA, after some delays, rejected the ICMSA complaints. The ICMSA also wrote to the the Pigs and Bacon Commission asking for the inclusion of millers and agricultural advisers on that board. The IFA considered that only producers and curers should be represented. The IFA Pigs Committee were unhappy that the ICMSA representatives on their joint committee were not pig producers and had no experience of the trade.[24]

The two fundamental causes of friction remained. No progress was being made towards an effective unity which could speak with one voice for all farmers. The ICMSA asserted its position as the established, independent representative of creamery suppliers. The IFA, even if their members supplied creamery milk, should not have Dairy Committees in each county; the IFA should not be involved in the National Dairy Council;[25] should not be active in raising share capital for Avonmore (or any) Creamery. In particular the ICMSA could not approve the collection of subscriptions for the IFA through the creameries. They wanted independent funding and equality between the organisations without any obligation to consensus or agreement on policy, under alternating chairmanship.

At the ICMSA annual general meeting in March 1978 James O'Keeffe expressed his impatience with the lack of progress towards farmer unity, making a strong plea to the IFA that they take constructive action. Otherwise the ICMSA would be forced to seek association with other groups to build a fully representative body.[26]

Anthony Leddy, of Cavan, succeeded James O'Keeffe on the unanimous vote of the ICMSA National Council. O'Keeffe had worked hard for unity of the organisations. Now Leddy made it clear that he would continue the effort in August 1978: If unity is the best way of servicing farmers, we won't stand in the way but it would be wrong for other pressures to force us together before the time is opportune.' He said that a united milk policy had been agreed with the IFA and co-operation was already functioning in four other areas, European affairs, taxation, animal health and rural development. The full range of agreed policies should be available for the autumn meetings of the two organisations. 'Whatever sort of organisation structure emerges from the present talks, my thinking would be that it must be aimed at achieving a better service for our farmers. There is a tremendous demand at present for proper representation of farmers at all levels and the present talks must bring about some form of overall good.'

Relations between Paddy Lane and Anthony Leddy were close and friendly. 'The pooling of IFA and ICMSA resources will be of immense benefit to the smaller farmer. A single Organisation will avoid duplication of services and provide new and more extensive services to farmers.'[27]

Agreement between the IFA and the ICMSA to form a Farmer's Union by 1981 was signed at Nenagh. A joint creamery levy of £1 per £1,000 would finance the new union, giving £500,000 a year from January 1979.[28]

The proposed agreement gave a special position to the creamery producers in the new body. This was to calm old suspicions of an IFA takeover. The two groups undertook to guarantee a basic income to Macra.

Long executive meetings in April 1980 reviewed the working of the joint committees. Alan Gillis and Des Kehoe argued that the ICMSA had got what they needed – information, working knowledge and back-up service, – without making any progress towards unity among farmers. In December 1981 reports on the working of the committees were submitted from both associations. The IFA, regarding the prime objective as the strengthening of the farmers' voice through unity, had gained little. The ICMSA assessment of the working of the agreement was even more directly critical and negative.

The end result of the IFA's invitation to co-operation in international representation was to introduce the division, duplication and wrangling of the milk sector to other commodities. The only possible benefit from organisational rivalry is that competition prevents slowth.

Policy Differences

No fundamental issue of dairy policy emerged between the organisations. Those which arose could have been avoided democratically if any one of several proposals had been adopted, such as the election of the creamery committee by the vote of all suppliers of milk to creameries, or the form of association chosen by the Beet Growers. There were a very few disagreements on practical policies.

*ICMSA favoured a 'Two-Tier' price system for creamery milk and for other commodities. Each farmer should get a high price for a basic quota of each commodity. The excess would be sold at a lower price. Thus small farmers would be favoured. IFA was not against the principle but, in agreement with the Department's committee,[29] considered the system unworkable. Under Neil Blaney the special price was given for the first 7,000 gallons of milk, but abandoned in 1972 under James Gibbons as Minister.

The NFA, with 80 per cent of its members small farmers, pressed for an overall small farm development plan. More in tune with European affairs, they would use the EEC structural policies, emphasising improved farming, with increased output and better marketing as a more constructive aid to all farmers than price increases on the home market.

Both agreed that, however regrettably, the very small farm could only provide an unacceptable income for slavish work. The IFA saw advantage in combining part-time farming with work in factory or agri-tourism.

*Disagreement on taxation was more apparent than real; on means rather than objectives. All accepted that farmers had to bear a fair share of tax. No farmer or small businessmen could willingly submit himself to the complications and costly of book-keeping under the existing income tax code. Income in agriculture is particularly hard to calculate on a yearly basis; it discriminates against family labour and against people with variable incomes. The ICMSA took a firm line, calculating taxable income from valuation of the farm. This would have the advantage of clear certainty, saving millions of pounds a year in book-keeping. However, it would not distribute the tax burden equitably between farmers, unduly favouring dairy farmers who have the highest incomes per Hectare or per £PLV. Agreement between organisations was reached at a meeting in February 1977, but not ratified.

Seeing no chance that the Government would accept and find workable a system of levies on produce sold or a full revaluation of land, the committee accepted, with the greatest misgivings, the fault-ridden income tax system under which the rest of the community suffers. The IFA had a victory in the abolition of rates (1982).

There was no difference in policy on capital or inheritance taxes. Farming, being heavily capital intensive because of the high price of land, would suffer more than the rest of the community under any property tax.

Both organisations withheld rates payment as a means of influencing Government.

Nitrogen Factory

The ICMSA favoured the Nitrogen factory project. The NFA considered the project doomed to costly failure and only ceased to oppose it on receipt of a guarantee that fertiliser prices would not be raised above that of imports.

Searching through the old reports, leaving out the harsh words used in controversy, there were only shades of difference on the best course of action. This is inevitable when any two groups work independently, in good faith, for a common purpose. In the joint commodity committees dealing with EEC matters there was little policy dispute. As in Irish national politics, the two main groups came to differ for the sake of showing they were different. The causes of the farmers' split were summarised as Pride, Prejudice, Personalities and Party Politics.

Pride

Prominence as an officer of voluntary organisations, such as co-operative committees or trade associations, confers status in the community and is a source of family pride. An effective public relations officer can ensure press coverage on any controversial issue. While no payment is usual, the social and business contacts; free travel to meetings; even trips abroad with conference entertainment, are a unique opportunity. To be quoted in the local paper confirms a man's position as spokesman for his neighbours. Any threatened loss of status through amalgamation of farm associations, local creameries or local government is naturally resented and opposed by committee men and managers.

An organisation develops a strong life of its own and an 'instinct' for survival. When two teams of delegates met they seemed to feel bound to best their opponents. Each side played tricks of bluff and one-upmanship which it would be as futile to record as it was harmful to their common interest.

Personality

Personality clashes accentuated the conflict, if they were not its original cause.

The first NFA committees included academics and those engaged in processing and marketing. Some of the farmers in the National Council had higher education or had served in one army or another. They were moulded by experiences outside the farm.

Juan Greene was a very large scale tillage farmer. He had a medical degree from TCD and held a war-time commission in the RAF before returning to the family farm. He believed that quiet reason prevailed better than confrontation and all men would co-operate when convinced by reasoned argument.

Rickard Deasy had studied economics and politics in Oxford. He had been captain in the Irish Army.

John Feely was of an older generation. An East Limerick dairy farmer, he had been active as a lieutenant in Intelligence during the War of Independence. After a dinner in Iveagh House to welcome a new German ambassador, he explained to a German and to Louis Smith that the present Government was not true heir to the founders who declared the Republic. He still had his gun for when the time came to revive the real republic. The German did not understand.

Charles Fletcher, with his sawmill and Jersey herd in Killeshandra, had little in common with John Feely's politics. He retired to the Isle of Man in 1971. He took rather little part in negotiations.

Later there was no such difference between the leaders. TJ Maher and Paddy Lane were southern dairy farmers having much in common with James O'Keeffe, Feely's successor. A National Council resolution was passed regretting the death of James O'Keeffe and appreciating the work he had done. TJ Maher and Sean Healy represented the IFA at his funeral, but there was still no concrete progress towards unity.

Politics

Neither association had political affiliation. Each excluded professional politicians from office. A majority of members of both probably voted Fianna Fáil. Those who saw all things with the eye of a politician labelled the IFA as Fine Gael; the ICMSA as Fianna Fáil. ICMSA became more radical in its policies over time: 'When Dan MacCarthy resigned as leader of the new Land League and joined ICMSA in 1982, this took the Association to the left of centre with a greater emphasis on land policy and farm restructuring as a basis for farm family viability'.[30]

Prejudice

After the first blows are struck in a quarrel, each side only remembers that they have been hit. The cause of the dispute is forgotten and only the enmity remains to blight even the next generation.

Finance

Organisation finance was a divisive issue. The ICMSA had a convenient, regular source of funds subscribed through the creameries. John Feeley's hottest words were against NFA encroachment on the creamery areas. Organisation was inexpensive, since little headquarters staff was needed. There was only one core commodity to be considered, others were peripheral. With few committees to run travel and research expenses were low. The ICMSA amassed a surplus which was prudently invested. Consequently the accounts of 1992 show that non-fee income covered two-thirds of costs. The association could survive even without members. There was a fear, shared in BGA and LMP, that the IFA proposed marriage with a view to 'the dowry'. IFA members of Council, Executive or commodity committees, were unhappy when they received less travelling expenses than their colleagues when attending the same meeting; the IFA had many calls for its services which it could hardly meet from existing income.

International Developments

The NFA was first in the field of international relations by a wide margin. The

ICMSA was at first lukewarm to the EEC (see chapter 8), but voted overwhelmingly in favour in 1972.[31] The NFA's enthusiasm dated from 1955.

When the ICMSA developed a more active interest in the prospect of EEC membership, continental leaders were approached with a view to membership of IFAP and COPA. The international bodies, having suffered the time-wasting squabbles of rival Italian organisations, would only accept one association from each country. The Organisation accepted should cater equally for all commodities and be recognised by the national government at home as the main farm Organisation in negotiations. Since the NFA was already established as Ireland's representative, the approach of James O'Keeffe and Jerome O'Shea to the Dutch president of COPA had to be unsuccessful.

The IFA could have exploited its exclusive right to international recognition in the struggle for membership at home. Its position in Ireland would automatically strengthen as the importance of decisions in the Commission in Brussels grew. It chose to combine forces with its rival to strengthen the farmers' representative position by presenting a unified front in negotiations – their long-term purpose from the beginning.

Even before entry to EEC it had become necessary to have a full time, resident, farmers' official in Brussels (Alan Dukes) in addition to the national embassy. It would have been clearly wasteful for the ICMSA to duplicate Irish farmers' representation.

The opening of the French market to Irish lamb illustrates the value of contacts in international committees. The first Common Agricultural Policy made no detailed provision for sheep, which were of minor importance to most Europeans. Thus France was free to protect its market against imported lamb. The Netherlands gained limited access, as described in chapter 7. Farmers' feeling in Ireland was so strong that the IFA deputy president, Joe Rea, feared the members' comments if he drove his French car to a meeting in Waterford. Sheep producers demonstrated in the streets of Dublin. Larry Lenehan, chairman of the sheep committees of the IFA and of COPA, Alan Dukes (in the Brussels office) and Lorcan Blake, chief economist, had long and friendly contacts with their French colleagues. They persuaded M. Clavel, of the Chambres de l'Agriculture, M. Monge, director general of the French Féderation Nationale Ovine, and his colleagues M. L. du Buveis and Count Henri de la Salle that a bilateral agreement admitting an Irish quota to the French market would be just and to their mutual advantage. M. Chirac, Minister for Agriculture and later President, was approached and agreement reached for an Irish quota of 39,000 tonnes.[37]

Then the IFA made the mistake of releasing the news before telling the Irish Minister. The Minister telephoned Larry Lenehan personally: 'What is in it for me if the IFA has taken the credit?' The matter proceeded no further. When James Gibbons became Minister, negotiations were successfully resumed. It was a matter of some comment by the NFU and the Scottish union that they were unable to gain similar access. In 1984 the Common Policy for sheep regulated the whole market.

Need of a Unified Association

The fundamental difference in opinion was the value of a dedicated single commodity organisation against that of a unified body. Irish family farms produced many products, specialist farms were very rare Perhaps, if Irish farms become highly specialised, commodity groups will be the right answer.

A loose federation of independent associations was the objective of the ICMSA. The central body would be like the Irish Congress of Trade Unions.[38] Each farm organisation would retain its identity, make its policy, manage its finance. The central body would co-ordinate policy but not bind its members. The NFA aimed at having one organisation to speak for farmers. Within that organisation commodity committees would work, co-ordinating their policies together and subject to the over-all National Council. Existing commodity associations could fit into this structure, electing their representatives to the commodity committee and to the National Council.

A workman's trade union has a straightforward task representing one skill with an employer. The Irish Congress of Trade Unions has done useful work but has not eliminated inter-union and demarcation disputes. A farm organisation is more complex. Few Irish farmers rely on a single product. Either the farmer would have to belong to a number of unions or request each commodity association to represent him in a full range of products. At local level this would be impracticable; at national level the Ministers for Agriculture would continue to complain that, to consult farm opinion, they must deal with the contradictory opinions of many farm associations. Sean Lemass, as Taoiseach, and James Dillon and Paddy Smith, as Ministers, complained of the impossibility of consulting farmers when a hundred associations purported to speak for them.

Forty years ago the need for a unified national body was clear. Development since then in Denmark, as in Ireland, is towards high technology and specialised farms, which have little common ground with traditional farmers. The tendency is for such groups to handle their own problems with increasing independence. Though ICMSA membership is not of this type, farmers' unions may be evolving towards more independent committees.

Stalemate

Both the ICMSA and the IFA lost the contest between farm organisations. More correctly the Irish farmers wasted time and money. The two organisations became more alike. ICMSA graduated from being a single product group agitating for milk prices to being a spokesman on all farm topics. It lost its position as sole spokesman on manufacturing milk, its special position with liquid milk groups, and failed to get full representation abroad.

The IFA failed in its main objective, of giving a unified voice to Irish farmers. On every topic there is a second, often discordant voice, listened to by those who dislike the first message. In 1966 the IFA felt forced to copy the successful aggressive methods of the ICMSA to the verge of demagoguery. There were demonstrations, marches, tax strikes, pickets on Government offices. Time, under happier circumstances, would have been better spent in productive work or planning.

Meantime, the Irish farmers pay twice for a weakened representation. Deputations are duplicated and time is consumed, though members and the public are unclear even on whether there is a practical difference between the rivals.

Work Resumed

The line between history and current affairs is not clear cut. The purpose of this history is to record events and people now almost forgotten. The events of more recent years, well remembered and served by journalists, can be treated more lightly. Up-to-date histories of Macra and co-operatives are in print. The ICMSA published their 40-year history in 1990.

Restored Relations with the Department

Committee work did not quite cease during the major dispute with the Ministers for Agriculture from 1966 to 1970. However, there was a break in the continuity of work. Neither time nor money were available for constructive planning. Restrictions on travel due to the Foot and Mouth outbreak in Britain inhibited

26 Discussions with the Minister for Agriculture, Jim Gibbons

meetings. Many projects had to be postponed because they required the partnership of the Department, co-ordinated through the Annual Reviews.

When the dispute ended there had to be a period of rebuilding, restoring the Association's finances, redirecting effort, re-organising staff.

From Cold War to Social Partnership

The immediate prospect of EEC membership presented a new situation. At national level farmers and Government could be seen to be on the same side. Controversial issues of price, grading, subventions were to be controlled and financed from Brussels rather than Merrion Street. The IFA's participation in 24 national committees was almost equaled by the number abroad. Foreign contacts forced closer relationship with the Irish Government. Regular weekly – often twice weekly- meetings took place between the Department, represented by secretary, deputy secretary, economist and heads of relevant sections and the IFA general secretary, assistant secretary, chief economist and relevant commodity committee representatives. Information was traded. The IFA contributed the regular reports on EEC developments given to IFAP by M. Rabot (the director of the Agricultural Directorate in Brussels.) and reported the COPA debates, in which it participated even before Ireland's membership in 1973. The civil servant who underlined in red the description of relations between Government and farm union in the Netherlands, 1954, was far-sighted (see chapter 6). The single Irish agricultural attaché in Brussels, however competent and willing, had not resources to cover the complexity of developments. He was heavily burdened in making travel arrangements and appointments for the numerous politicians and delegations visiting Brussels. The IFA, once again, found itself better informed than the Department, but this time in a productive partnership.

A rough measure of the increased involvement of the IFA in European issues is given by the pattern of expenditure. In 1993 49 per cent of expenditure was under this head. Farmers' reaction to national and European influences and the farmers' developments were closely connected. Unlike its sister organisations abroad, the IFA received no government assistance in representing the national interest after 1965.

During the honey-moon years, 1973 to 1978, the general rise in prices concealed, for the moment, the rocks ahead. Gross agricultural output rose 183 per cent in value and 31 per cent in volume; agricultural exports rose ten-fold in value; co-operatives were consolidated and revitalised.

Renewed tension

A number of factors contributed to a worsening of IFA relations with Government.

In the election of 1973 Fine Gael gained five rural seats, this, combined with Labour's gain in Kildare, Tipperary North and Waterford, cost Fianna Fáil its majority. Policy became more urban-oriented under a coalition government with Labour

The Irish Government, pressed to finance its extravagant expenditure, attempted to claw back some of agriculture's EEC gains by Richie Ryan's 'Incentive Budget' of 1977. As Paddy Lane put it: 'It looks that it is the policy of the Minister for Finance to negate all the gains of Mark Clinton for farmers. For example the recent British package gives 25 pence a gallon increase in milk, but already the January Budget, by changes and increases, has taken over 20per cent of the milk price increase.'[1]

The Coalition's 'Incentive Budget' reduced taxes on industry. The good year of 1976 attracted attention. Farm taxation increased, in spite of the NESC report on farmers' income tax, which acknowledged that farmers subject to income tax were paying their share share of tax twice, in rates and again in income tax. The tax threshold was reduced from £100 PLV to £75 PLV; the multiplier applied to PLV to estimate taxable income was raised from 40 to 65. Costs formerly allowable against tax, such as wages, contractor charges, interest over £2,000 p.a. and depreciation on farm machinery, were disallowed. The effect was a doubling of assessed income. The farmer of £80 PLV, free of income tax in 1976, was now liable for £958; tax on a farm of £159 PLV was raised from £375 to £2,846. Rates, VAT and other taxes were also increased. Efforts were made by the Valuation Office to tax pig houses as factories instead of as farm buildings (September 1977).

Relations between the IFA and the Fine Gael/Labour Government deteriorated. The Exchequer foreign debt rose from £126 million (1973) to £1,040 million (1976) during Richie Ryan's period as Minister for Finance.

The government, seeing agricultural prosperity, cut the services and supports of the Department of Agriculture. The IFA president published the 'Deasy Decline Index', so called after the Minister, Austin Deasy. The justifiable reaction of Government in reducing farm supports and increasing taxation with rising income impacted on farming just when incomes began to fall steeply. Secondly, the land tax (1984) was seen as inequitable and divisive of farmers. Thirdly, the exchange rate changes, leading to the MCA problems described elsewhere, were harmful to agricultural and other food exports. A couple of bad harvests later, the shadow of CAP reform worsened the farmers' temper.

Elections 1977

In preparation for the General Election of 1977 the IFA prepared a questionnaire for candidates, asking their policy on farming issues. Fianna Fáil were quick to exploit the opening presented by Richie Ryan's Budget proposals, taking two pages of the *Farmer's Journal* (16 July 1977). Among many items were:

- Income tax assessed on the 'Notional' basis instead of on accounts to be retained as an option.

- Rates on agricultural land to be allowed as an instalment of the farmers' tax bill.
- Tax to be based on average income over three years, with benefits for investment, allowance of wages (relief value £2.5 million per annum) and payments to contractors (£1.5 million.).
- Abolition of the tax on co-operatives, introduced in 1976.
- Review of wealth and capital gains taxes, to exempt long term gains and the transfer of farms to sons.
- Pensions for farmers and a Farm Inheritance Counselling Service.
- Loans at three per cent to developing farmers.
- Full compensation for cattle culled for disease.

Fianna Fáil were returned to power with a large swing in rural areas. Paddy Lane, IFA president, commented in the *Farmer's Journal*:

> Very many farmers have changed their allegiance because of the agricultural policies as promised in reply to the (IFA) questionnaire. I believe Fianna Fáil, if they are to remain a creditable party, must implement all of these policies in the immediate future.
>
> I would ask farmers to hold on to these two pages (of promises in the IFJ) for the next four years and compare the performance of Fianna Fáil in office with the promises of that party prior to the General Election.
>
> As farmers have become part of a floating vote and by doing so have been instrumental in getting Fianna Fáil into power, equally so the same thing can happen in reverse. Thus, even though the number of farmers is declining, the political strength of farmers in the years ahead will be more crucial than ever before.

These promises were attractive to farmers but were criticised as failing to stimulate growth in this time of opportunity. The IFA looked abroad and saw that competing farmers had better conditions. In New Zealand, for example, all farmers were liable for income tax but the incentives to growth were clear. There was a tax credit for every additional livestock unit. An extra sheep gave £16 tax credit; a cow counted as six sheep. In effect stock could be bought from pretax income. The full value of the stock could be written off against tax in the year of purchase – a concession common for European and Irish industrialists. An expanding farmer could postpone tax indefinitely. When mature increased income would benefit the State with full taxes. Taxes became due when the farm was sold, but if the farmer was retiring or passing the farm to a son, tax concessions were carried forward. To accommodate the ups and downs of farm income tax in a good year could be carried forward in a special fund and drawn upon in a bad year. Local rates were for roads at a level of 70 pence a productive acre.[2]

Fianna Fáil won the election.

The new Minister, Jim Gibbons, stated in the Dáil his intention to make his party's manifesto his Bible for the next few years. He promised fiscal reforms – improved VAT concessions; the notional income assessment would continue indefinitely. Co-operatives were the ideal instrument for controlling the advisory services. Pending a take-over by the co-operatives, County Committees for Agriculture would cease to be appointed by the County Councils. They would be elected by the votes of the food industry, farmers and co-ops.

The Manifesto promises were not fulfilled. Farm income fell from £820 million in 1978 to £672 million in 1980. Farmers held protest demonstrations in Kilkenny and Limerick. The fall in the value of the Irish pound eroded the benefit of the Common Market because Intervention prices were set at the old 'green' rate till 1983. Bad weather in 1980 and 1981 made matters worse.

The justifiable in reducing farm supports with rising incomes impacted just when incomes began to fall steeply.

The IFA placed a 'Survival Strategy for Farm Families' before the Taoiseach on 26 February 1986. Main elements were: devaluation of the Irish pound by eight to nine per cent. This was done on 2 August. Then the IFA campaigned for a similar Green Pound devaluation. The Council of Ministers Brussels conceded a devaluation of 5.8 per cent for animal products and 2.8 per cent for grain, in September. Farm prices increased accordingly. Irish concessions in loan schemes and an undertaking that cuts in government spending on agriculture be no greater than in other sectors were sufficient to cause the National Council meeting of June 1986 to defer its decision to call for the resignation of the Minister, Austin Deasy. A significant recovery in farm income was evident in 1987 and 1988.

Farmers Recognised as Social Partners

Government had consulted with the trade unions and the employers' associations in various National Wage Agreements, the National Understanding and other developments from which the IFA was pointedly excluded.

When Fianna Fáil returned to office in 1987, Joe Rea, IFA president, had established good relations with Charles Haughey, now Taoiseach. He told the National Council (March 1987) that the Fianna Fáil election manifesto had emphasised the importance of agriculture and the food industry in national recovery, but had been remarkably short on specific proposals. The onus was on IFA to make concrete proposals. The result was a policy document 'National Recovery – the Role of Agriculture and Food Production', presented to the Taoiseach at the National Council Meeting in April with the reminder, 'both parties have gained in wisdom and understanding the intervening 20 years' since Mr Haughey's last address to the IFA. The Taoiseach commended the constructive role of the IFA in the preparation of the National Economic and

Social Council's report on 'A Strategy for Development' the previous November.

Relations with the Government were now on a friendly basis. Through the remainder of 1987 the IFA worked with Government and the social partners in preparation of the 'Programme for National Recovery'. The National Council gave it unanimous endorsement in October. The ICMSA, ICOS and Macra were also involved in the negotiations. However, the ICMSA withdrew its support because the report did not accept their recommendations on taxation.

The Programme was successful. Government expenditure and borrowing were curtailed; interest rate fell and inflation fell to two per cent in 1988. Farm incomes rose 16 per cent.

Farm organisations continued to participate in future National Agreements. The IFA does not take a high profile in the national planning bodies (NESC, PNR, PESP). The focus is on Brussels and the Department of Agriculture. One observer,[3] quotes trade union opinion: 'It's really ourselves, the FUE and the Government doing the negotiating: we decide on the Programme, and then see what grants we'll give the farmers, and then we try as hard as we can to get them to pay a bit more tax.' Again: 'The farmers' only weapon is their vote, their numbers are falling, and their money comes from Brussels anyway. Even allowing that each farmer influences family votes and those of linked industries their electoral weight will become small.'

WORK OF COMMITTEES

Taxation

Taxation became an acute concern for Irish citizens as government expenditure rose. The voters, who had allowed themselves to be bought by extravagant benefits from Government, were angered as taxpayers when the bill was presented. Taxation as a share of net national production rose from 38.7 per cent in 1975 to 39.3 per cent in 1979 and 57.4 per cent in 1985. It was estimated that in 1979 one million people protested in street demonstrations against the inequities of taxation.[4]

Nothing caused such resentment as income tax. Inflation had been used to increase this tax. Inadequate allowances for the fall in the value of money brought incomes of social welfare level into the income tax net; 40 per cent of workers became subject to income tax at the high rates originally designed for the rich. Naturally the PAYE workers resisted and looked round for tax dodgers to relieve their burden.

As in the 1950s, the urban workers perception was that 'farmers pay no tax' – meaning they did not pay income tax on the same basis as city workers. Rates, the farmers' main tax, were ignored. When Common Market export opportunities raised the real farm income by 70 per cent farmers' prosperity attracted attention. It was little noticed when, in 1980, real incomes fell 50 per cent, back to the starting level of 1970.[5]

The points at issue were:

1) Distribution of tax between sectors of the community.

2) Basis of assessment.

Distribution of Tax

The IFA had always insisted that farmers were willing to bear their fair share of tax. A sharp difference of opinion arose with outsiders on what was a fair share. Farmers also looked at the taxes paid by their competitors abroad and argued that Irish farmers'competitive position by their heavier burden. In 1976, only 7,000 out of two million French farmers had to present accounts for income tax.

From its first year the NFA continually pressed to have rates recognised as the farmers' income tax. the abolition of rates on dwelling houses, an election ploy in 1977, made it clear that the tax was levied on the means of production, land and business premises. Alan Dukes, IFA economist, made a detailed comparison of the tax burden of farmers under rates and income tax, based on the National Farm Survey 1966/67. The broad conclusion was that a system of taxation based on actual incomes would achieve a much fairer distribution of the tax burden between farmers. Small farmers would, like low wage-earners, pay no direct tax. Only farmers with over 200 acres; bachelors and liquid milk producers, would pay more if subject to income tax instead of rates.

Method of assessment

Income tax, calculated as a multiplier of poor law valuation, was introduced in 1974 as an addition to rates on land. Next year the Government and the IFA agreed to a review of farmer taxation under the auspices of the National Economic and Social Council.

The IFA submission, accepting that the farmer must pay his fair share of central and local tax, pointed to the multiplicity of taxes involved. Sensitive to the accusation that they 'paid no income tax' , the farmers wanted rates to be combined with income tax and given same earned income and other reliefs. They wanted income to be averaged over three or five years, as in New Zealand. (Incomes which vary widely attract additional tax. Reliefs are not fully used in bad years, while more income enters the high tax level in good.) Farmers should not be placed at a disadvantage with their competitors abroad by higher taxation. As Ireland's main exporters, they merited the 10 per cent rate of profit tax given to exporting industrialists.

The NESC report was very competent. The matter was again examined, with less appreciation of the problems, in the 4th Report of the Commission on Taxation. Little heed was paid to any of this work by the authorities.

No one would voluntarily face the complications of the unreformed Irish income tax code applied to the non-farming citizen. The system was developed in an English industrial environment where salaried income is easily assessed. It is ill adapted to family business, where ownership and the earnings of unpaid family labour belong to the group, not to the head of the household. Farm accounts are complicated by the many products and many small transactions involved. The income from a beast, sold at two years old, cannot be determined neatly on a yearly basis.

The cost of account keeping exceeds the tax revenue on small farms. Most had incomes below the threshold of income tax liability. The alternative system was income tax on a 'notional' basis. That is, on the probable income, based on land quality or type of farming. This is simple and cheap to assess. However, the notional system based on poor law valuations of 1856, used for rates and the new income tax, bore no relation to reality, as the NFA Rates Committee had pointed out in the 1950s and as the Supreme court judged in 1982. Such a system, based on a crude measure, taking no account of annual variations in overall farm income; of individual farms; of varying fortunes of tillage, beef or milk production could not be just. Farmers feared that income tax would be imposed in addition to rates and, once imposed, it would become permanent and subject to indefinite increase, regardless of promises. The notional farm income for income tax purposes was raised to 65 times PLV in 1975. In succeeding budgets the multiplier was increased to 95 times PLV and 125 times in 1979; the tax on co-operatives was continued; the County Committees remained in charge of the advisory services; rates were raised; farm incomes fell; a levy of two per cent was imposed on sales of cattle, milk, pigs, sheep, sugar beet and cereals was designed to recover part of the cost of services such as disease eradication.

By October the 7,000 farmers newly netted by income tax were finding that the 'notional assessments' overstated their incomes. Even when they appealed successfully they had to pay half the assessment within two months, pay an accountant and wait for the return of the overcharge.

IFA Tax Policy

Any change inevitably favoured some members at the expense of others. Some counties favoured an income tax because it would vary with ability to pay and, being seen to be like other workers, would lessen the tension between town and country. A fixed rate would attract tax even in loss-making years. On the other hand, Waterford County Executive wrote to the general secretary demanding the dismissal of Seamus MacCann, secretary of the National Taxation Committee. He had stated in Waterford before the Committee chairman D.J. Coughlan that farmers must expect public pressure for the imposition of income tax and should prepare their case (1968).

Within IFA a consensus opinion in favour of income tax based on farm accounts emerged as incomes fell. D.J. Coughlan, chairman, reported the findings of the Rates and Taxation Committee to the National Council: '1) The taxation system for farmers should be income tax based either on accounts or notional. 2) The present accounts system with adjustments (such as stock valuation revision and 100 per cent depreciation of buildings) is suited to farmers generally. The notional system should be retained but needed reform.i.e.PLV would have to be ironed out.' The president, Donie Cashman, summarised the debate. The Council were: 'in favour of income tax as being the only fair system of taxing farmers and IFA would be prepared to negotiate for income tax for farmers. He stated that the farmers do not want rates or resource tax.'[6]

Joint Tax Committee IFA/ICMSA

IFA and ICMSA attended a meeting, requested by the Revenue, to discuss the Revenue's proposal for a voluntary 'farm profile form', for submission by low income farmers claiming Income Tax exemption. Revenue would not entertain the ICMSA's 'production unit accounting system' as an alternative to accounts. It was also stated to the farmers that reform of the Poor Law Valuation system would take eight years and cost £10 million.[7]

The ICMSA then withdrew and refused to attend the next meeting. The Chairman of the NFA Tax Committee reported to the National Council that a fruitful discussion on the farm profile system took place at the second meeting with Revenue but, in the absence of the ICMSA, no decision could be reached.

By the time of the 1979 Budget IFA/ICMSA unity talks were at a relatively advanced stage following the 'Nenagh Agreement' (see chapter 13). Two joint Special National Council meetings discussed the budget (26 February and 1 March 1979). Between the meetings the two organisations met the Taoiseach, Jack Lynch, and agreed that: 'the Government will consult immediately with representatives of the farming organisations and will devise a system of income taxation for farmers that will bring in a yield from them in line with other sections of the community.'

The 'joint policy development committee' met officials of the Department of Finance on six or seven occasions, but unexpectedly terminated its meetings on 29 March.

The Special Taxation Committee of representatives of the two farm associations met from 1979 till 1985 in unsuccessful attempts to agree a tax policy.

The chairman, Donie Cashman, explained that he had himself proposed a system of tax based on 'adjusted acres', but was now convinced that it could not satisfy the criteria set down by the farmers. The views of 16 of the 19 present are recorded in the minutes. With the exception of D.J. Coughlin, of Waterford, all favoured the proposals and stressed that accounts were the only fair and feasible basis of assessment. This opinion was shared by the three ICMSA representatives: Tom O'Dwyer, P. Hourigan and T. Ryan.

A copy of the report was sent to both National Councils. The IFA adopted the report as national policy at the September Council Meeting.

From that point on sharp disagreement developed.

The ICMSA made the valid point that accounts for all farmers would be most onerous, costing £230 a farm or £27.6 million a year. This would be wasted money for the majority of farmers with income below the tax threshold. They proposed 'a Production Unit Accounting System' based on the farm management survey. This system, based on average incomes, cost little to asses. The system would stimulate production and efficiency, since profits in excess of the average would be tax-free.[8]

The IFA replied that the system would not be easy to administer; farmers with an actual income below the tax threshold in a given year would be forced to pay excessive tax or hire an accountant to prove loss.

The discussions at the Joint Committee and outside became acrimonious. The ICMSA accused the IFA of departing from the agreed policy and demanded full support for the ICMSA 'production unit' system of assessment. Members were advised not to fill in the Farm Profile form. The IFA president, Joe Rea, when pressed to reply in kind said, 'ICMSA have done everything possible to get me to kick them, but I have not and I will not because a public media controversy between farmers or farm organisations at this time would only ensure the success and speedy implementation of a Land Tax against all our wishes.'[9]

Since farm representatives disagreed they were both ignored. The 1979 budget of George Colley, Fianna Fáil Minister for Finance, announce major increases in farm taxation. Income tax was extended to smaller farms, the threshold was lowered from £60 PLV to £70. The presumed income was increased to 125 times PLV. A levy of two per cent on all sales of cattle, milk, pigs, sheep, sugar beet, introduced to pay for services such as disease eradication, caused the greatest concern. The Minister gave notice of future tax increases. This was the year when farm income began to decline.

With hindsight, it is clear that the government decisions of 1979 bedevilled the farm tax issue, leading to a confrontational relationship instead of the negotiated approach between the IFA and Government which was developing.

In the 1980 budget the notional system of assessment was abolished, the income tax threshold was reduced to £40 PLV but a simplified form of accounts was introduced for farmers entering the tax system for the first time. In addition a resource tax on all holdings over £70 PLV was introduced.

Other Taxes

Other taxes multiplied in response to the Government's hunger. IFA had to state the farmers' case on many taxes:

Value Added Tax presented special problems in agriculture. It is costly to collect from small business. In Britain, with roughly the same system, the costs of operating the tax averaged 29.6 per cent of the revenue collected and were three times greater than revenue from shops with turnover under £10,000. (1977/8).[10] The IFA eventually got the rate reduced from 10 per cent to three per cent in 1981.

The absence of wealth tax was widely criticised on 1974. It was useless for the NESC report on Farm Taxation to show that Irish capital taxes were, with Britain, the highest in the European Community – three times those in France, Italy, Netherlands, Denmark; eight times those in Germany [p. 126]. Wealth taxes yielded a tiny part of revenue in other EC countries, ranging from nil in Germany to 0.9 per cent in Denmark. The issue was seen as important by those in politics who set equality (not to be confused with equity) as their prime objective. As applied to farmers it was a most inequitable burden.

The IFA resisted wealth taxes as a selective attack on farmers. Though earning less than an industrial worker they have more capital in land, stock and equipment per person employed than most heavy industry. The tax was imposed in 1973. The IFA argued that the rate of 1.5 per cent to 2.5 per cent was penal when the return on agricultural capital (including land) was 2 per cent.

Farmers, it is said, live poor and die rich.

Capital taxes became an acute problem on medium-sized farms when land prices at auction rose tenfold after entry to EEC.

(The burden was postponed by evasion. Farmers did not bother to take out probate, legally transferring land from father to son. If Paddy Reilly was succeeded by his son Paddy and no one disputed the will, why bother to pay taxes and legal charges? Some 40 per cent of farms had defective title when bank borrowing became common in the 1970s.)

The selective impact of wealth tax irked the farmers. Pension funds, though a huge investment, were never imputed as part of the urban employee's wealth for tax; personal belongings like jewellery or paintings, though unproductive, were not assessed or could be concealed; much investment was through tax havens. Productive investment in agriculture or other family firms was penalised because it could not be concealed.

Farmers suffered from the general bias of the tax system against the self employed family firm. Efforts to make common cause with other small business people (1975) were not fruitful.

Death duties are onerous on family business. They are unpredictable, imposed when the 'managing director', the key man of the labour force, is removed. The payment is made when legacies to the members of family who have left agriculture become due. The whole value of the farm is attributed to the deceased, though the investment in the farm represents the accumulated savings of unpaid family labour. A legacy is often the payment for years of unpaid family labour.

Prior to 1970 only 90 farmers a year actually paid estate duty. As with other taxes, ever poorer people became liable because tax thresholds were not adjusted for inflation. The rising value placed on land endangered the family farm.

Death duties were removed but other inheritance taxes took their place. It became difficult to pass on a medium-sized, viable family farm to the next generation. Consideration was given to the incorporation of farms as companies and various forms of inheritance but none proved effective.

Similar problems arose with capital acquisitions tax, in effect a form of inheritance tax.

The resource tax on farms over £100 PLV was contested for years. The IFA took the case to the High Court in Autumn 1980. Eventually a refund of £700,000 was obtained in 1982.

PLV Action Committee

The 23rd of July 1982 was a red-letter day for farmers. Mr Justice Barrington, after 14 days' hearing in the High Court, found in favour of five Wexford farmers, members of IFA and of the PLV Action Committee, in their plea that Rates based on poor law valuation were contrary to Article 40.3 of the Constitution. The farmers argued that the valuations were 'arbitrary, discriminatory, unequal and inequitable basis for the imposition upon the plaintives of taxes ... and for the assessment of their eligibility for benefits.'

On appeal by the Attorney General the Supreme Court quoted with approval Barrington's judgement that the valuation system was

> shot through with unnecessary anomalies and inconsistencies ... The object of the Act of 1852 was to make one uniform valuation of lands and tenements which may be used for all public and local assessments and other rating.' ... in relation to land it was provided that such valuation 'shall be made upon an estimate of the net annual value thereof with reference to the average price of several articles of agricultural produce hereinafter specified ... The prices were the average prices which obtained in the three preceding years – the years immediately following the Famine ...
>
> In the course of this evidence it was recalled that the Griffith Valuation upon which the present system of land valuation is based, took 14 years to complete. It commenced in 1852 at a time when the widespread agricultural depression which followed the Famine was apparent throughout the land. It was completed 14 years later, at a time of comparative prosperity which, although they were bound by an index of agricultural prices, must have caused the valuers to take a better or more optimistic view of land values. Artificial fertilisers were then unknown and alluvial soils which were chemically well endowed and naturally fertile were regarded as prime agricultural lands. These soils, although inclined to softness, were easily worked by spades on the ridge and furrow system. Today with the development of fertilisers and agricultural machinery taking the place of the cheap labour of other days, it

is the higher and drier land which is prized as productive, while alluvial soils are regarded as of little value ...

The court is satisfied that these conclusions or findings are fully supported by the evidence ... and for that reason cannot be upset on appeal:

1) The existing valuation system does not provide a uniform system for valuing lands throughout the State.

2) There is no consistency between county and county or within individual counties.

3) The valuation system has failed to reflect changing patterns of agriculture with the result that the land which modern agriculturalists would regard as good land often carries a low valuation, while land which modern agriculturalists would regard as inferior often carries a higher valuation.

4) The whole system is shot through with unnecessary anomalies and inconsistencies ...

The result is that the five plaintiffs, although their land is of poor quality , are made liable for rates at a level higher than their neighbours have to pay for significantly better land ...

In the assessment of a tax such as a country rate, reasonable uniformity of valuation appears essential to justice. If such reasonable uniformity is lacking the inevitable result will be that some ratepayer is required to pay more than his fair share ought to be. This necessarily involves an attack upon his property rights which by definition becomes unjust.'

The effects of Barrington's judgement were immediate. No rates had to be paid. Any outstanding rates due need not be paid. Any qualifications expressed in terms of PLV for health or other benefits no longer applied.

Had the State carried out regular revaluations, as was provided for under the original Act and as is done in other countries, this case would never have arisen. Slovenly administration by the State and a disregard for justice had led to overtaxation of many citizens for a hundred years. The NFA had drawn attention to these and other anomalies continually for thirty years. This single victory justified the existence of the IFA in cash value to farmers and in equity between taxpayers. Unfortunately it also destroyed the slender basis of responsibility and independence of local government.

Taxation of co-operatives presents a difficult tax problem. The individual member is usually of modest income, hardly reaching the tax threshold. A co-operative does not make a 'profit'. At the end of a year it hopes to show a 'surplus', which is then put before the members to distribute at their general meeting , if they see fit, as a bonus on trade, not as a dividend on capital. They cannot make a profit from trading with themselves. Some of this surplus they will normally reinvest for development of the enterprise, or decide to pay out at some later date. Since the co-operative is not separate from its members, the operating surplus remains the property of the members trading with the co-op and should be treated as part of the members' taxable income.

If, following prudence and in accordance with Government policy, a number of small farmers come together in a co-operative, their combined enterprise should not be taxed more heavily than the sum of their individual liabilities.[11] On this reasoning all co-operatives had been free of income taxes till 1966. Logically, trade with non members is fully taxable.

Competing private merchants complained that this was an unfair competitive advantage.

In 1976 the special tax position of co-operatives was revoked. The IAOS was particularly involved, but the IFA and the ICMSA also protested. It is not evident that the Government was especially concerned with abstract theory or ethics. The coops had become large and financially sound. A government reduced to borrowing 13 per cent of Gross Domestic Product to meet its extravagance could not be expected to forego an easy source of funds. The promises of Jim Gibbons on becoming Minister for Agriculture in 1977 had never much prospect of fulfilment.

Removing the tax advantages of co-operatives also removed the impediments to the formation of PLC's, such as Waterford or Kerry. These large firms, with enterprising management and the newly found access to investment capital, became much more active, efficient competitors. The private businessmen had 'sown dragons teeth' and now met fiercer competitors.

Industrial Committee

The Industrial Committee continued and expanded its work at the iterface between farm and other business.

Input Quality

Price was the first complaint of farmers when they took the fertiliser manufacturers to the Prices Commission. Competition proved a more efficient way to bring prices down. Prices quoted by co-operatives and merchants through the country were surveyed regularly. Published in the *Journal*, these enabled farmers to shop at the best national prices. The country became one market with prices known to all. Agriline carried the latest information on current prices to the computers of all subscribing farmers from 1988.

Value for money, equally important but harder to document, became the next preoccupation.

Fertilisers

The six manufacturers met with the IFA twice yearly to consider prices and quality. Important findings led to savings of many millions of pounds. For example, of 14 brands examined nine failed to reach manufacturers' specifications in the three compounds tested.[12] All main suppliers were covered. Two cases did not involve the farmer in any cash loss. In the other seven, excess cost varied from £1.43 to £9.69 per ton, averaging £6 (1982). Similar surveys were published regularly in the *Farmers' Journal.*

Compound Feed

A similar range of tests showed defects in compound feeds: In 28 samples of dairy ration six were outside the legal tolerance for crude fibre. All but three had excess ash. Similar rations showed a price variation of £34 per tonne. In 11 samples of Calf Ration five were outside the legal tolerance in crude fibre and two in protein. The price variation was £30 per tonne. All 10 samples of Pig Fattening Rations had excess moisture; six low lysine reading. The price variation was £19 per tonne. All except one of the 19 samples of Sow Ration had excess moisture; four low lysine reading. The price variation was £30. Only one ration was found correct in all specifications.[13]

On the annual fertiliser expenditure of £350 million a saving of £1 per tonne increased net farm income by £1 million a year.

Quality of lime was similarly tested from 1989.

Quality of concrete readymix, plastic covers for silage, new roofing materials, sheet iron, farm machinery were also tested

Way Leaves

Payment for wayleaves by ESB and the Department of Post and Telegraphs, brought £40,000 a year for 20,000 wooden poles on farmers' lands a further £50,000 was obtained for a realignment of main lines. Periodical increases to compensate for inflation were negotiated (35 per cent increase in 1981). Substantial sums were involved in other wayleaves. The Cork-Dublin gas line paid £150,000 a year, reviewable every five years. After some litigation waterworks paid wayleaves, for example, 21 Monaghan farmers received £200.000 for waterworks. Litigation was successful against Dundalk Waterworks and Louth County Council in 1988.

In 1985 alone there were 70 meetings with local groups; agreements were reached with ESB, Bord Gais; County Councils. A Wayleave Manual on farm rights was published.

The IFA charged members one per cent of the revenue obtained for this special service.

Live Stock and Disease Eradication

The vets strike, lasting 21 months to October 1976, caused a major setback to disease eradication. This was one further illustration of the legal flaw by which any trade union can unite to refuse service, regardless of the damage done to the community and without question of the seriousness or frivolity of the dispute. By contrast a 'conspiracy in restraint of trade' by any group of the self employed, such as farmers, is a criminal offence.

In eradication of Bovine Tuberculosis and Brucellosis the Joint IFA/ICMSA Committee had to express dissatisfaction with progress to 1979. TB had been eradicated in six EEC countries but still affected 2.92 per cent of Irish herds and 0.19 per cent of animals – not a great improvement considering the expenditure of £700 million between 1954 and 1980/81.[14] The cost is now over £billion and the problem seems insoluble with present methods. The Committee expressed concern each year at inadequate compensations and financial cutbacks.

More than 25 per cent of herds in the dairying South were still infected with Brucellosis. Kildare was worst with 43 per cent. A substantial improvement has been made since introduction of the preintensive eradication scheme. Infected herds fell from 43 per cent in 1973 to 27 per cent in 1978.[15] The 'final eradication campaign' was started in 1984.

Warble-fly eradication was concluded more rapidly than in Britain or France, in spite of some hitches. Unsuccessful before the War, the effort was restarted by the organised farmers of Donegal. Between 1965 and 1976 it involved 32 million treatments. An increase in the counties Kerry, Clare, Longford and West Meath called for a new drive in 1984.

The Cattle Advisory Committee, introduced by the Department of Agriculture in 1978, was considered an attempt to control breeding on the old disastrous lines. It was 'composed of 17 bureaucrats, retired farmers, beef farming interests, with no more than two or three with any commitment to present day commercial dairy farming.'[16] Mark Clinton, as Minister for Agriculture, refused to meet the IFA to discuss animal health, reactor prices, liquid milk prices on the ground that the Animal Health Council was charged with these matters. The system of democratic election of farmer representation was not yet fully accepted.

Modern methods of stock improvement took over as was confirmed by the final removal of the visual bull licensing scheme in 1983.

Marts

A new agreement was made with the Banks Standing Committee to limit spread of marts. Shareholding requirement was raised by £10,000 to £50,000 for new marts and £30,000 for existing co-ops. The catchment area must be 10 miles radius and 20 miles from any other mart. A minimum of 500 share-holders was specified for new co-operatives, a majority of whom must be full-time farmers. The general standard of premises must be approved.[17]

Live Cattle Export

There are a host or reasons why it is better to export meat than live cattle. Payment can be made for quality, transport cost should be less, employment is given. Traditionally importing countries protected their meat industry and would only accept Irish cattle live. Having lost the controlling influence of co-operatives, such as Clover Meats, farm organisations have taken the short-term view that exploitation by private slaughterers should be controlled by live export. A fundamental reform of the industry would be in their long-term interest.

Grain Committee

In the Grain Committee negotiations were continual. The abolition of the Grain Board, which had provided a structure to the market, was regretted.[18] Millers refused to reconsider their offer of £76 a tonne for unmillable wheat, though the IFA Committee pointed to co-ops and merchants in Waterford giving £90 a tonne. The week following IFA publicity the millers improved their price to £87.

Horticulture

Having failed to respond to the early warnings of increased competition, the horticulturalists faced difficulties. Grading and marketing was the Achilles' Heel. The early effort of North Dublin Growers' Co-operative foundered on disloyalty. Members sold to the co-op as a last resort. Grading was reluctantly raised in standard only when the imported product had gained a foot-hold on the home market.

A programme to capture the £27 million spent on replaceable imports was launched by ICOS and the IFA Committee in 1983. It was expected that home consumption would grow a further 5 per cent a year with increased consumer incomes and better presentation of produce. A national organisation should be organised on a county basis with specialised groups for: Field Vegetables, fresh Field Vegetables, processing; Glasshouse crops, edible; Glasshouse crops, non-edible; Mushrooms; Nursery stocks; Soft fruits; Top fruits.

Government promised £1 million for start up, of which £100,000 was provided in the 1984 Budget. At local level 12 co-operatives were formed, with a national federation.

Representation was established on the Irish Potato Marketing Board. A case against Ciba-Geigy for defective potato spray was won by 35 Meath farmers on behalf of all growers (1981).

In November 1985, spurred on by the destruction of the potato crop in the wettest harvest season on record, Commercial Horticultural Producers of Ireland was launched by P. Hegarty, Minister of State. Funded by growers, Government and the Dublin Industrial Estate in Glasnevin, it contracted with Quinnsworth supermarkets and had full backup from Teagasc and other services.

Farm Structures

Low income in farming is related to size. Good farming can lessen the problem. The Farm Apprenticeship Scheme was followed by many schemes of the Brussels Commission. The association's small farm proposals of 1966, contained in the Green Book of 1970; developed in the Low Incomes study of Western farms (published by IFA in 1978 with the assistance of AIB), showed one approach. The efforts to diversify into new products of agri-tourism, forestry, horses, deer and special crops, or industrial development in rural areas were all part of a general effort to stave off rural decay.

The Departmental Guide lines, proposing that grants be withheld from farmers with off-farm earnings over £3,500 in 1978, was opposed by the IFA Rural Development Committee. This is a frequent dilemma of farm politics. French farmers in the 1950s protested when film star beauty, Brigid Bardot, made grant application for her farm. They argued that, having physical assets they lacked, she needed no agricultural aid. In the small-farm areas there is little regular employment, the best hope of raising the standard of living is by part-time industry.[19]

Agricultural Relations Board

As agriculture's share of national income and employment declined its importance became less obvious to the non-farming community. Good relations had to be established with the urban population. The Consumers' Association was important as representing the farmers' chief customers.

When a heart attack forced Sean Healy to retire as general secretary in 1977 he took on the supervision of Rural-Urban Relations Committees, established under the auspices of FBD and the *Irish Farmers' Journal*. There were 22 committees in 1983. Communications courses were arranged for branch leaders, with particular emphasis on the local radio stations which were becoming influential.

Welfare of farmers' children in third level education was aided by co-operation with suburban women's clubs, providing accommodation.

Land Reform

The Government land reform proposals of 1977 introduced long term leasing of land in an effort to make land available to developing farmers was welcomed as, 'fulfilling the primary aim of the Rural Development Section of IFA for two decades'. Erskin Childers had made the same proposals when Minister for Lands. The basic flaw was that the owner saw little or no chance of recovering the property at the end of the lease, even if the law changed.

Macra was less enthusiastic. They said that rented farms were only one per cent of holdings and unlikely to become important.

Farm Liability

Common Law placed surprising liability on farmers for injury, even to unwelcome trespassers. Pressure was applied to change the law and agreements were made with gun clubs and hunts for insurance. A few horsemen do little damage to permanent pastures. The increased popularity of hunting brought a hundred horsemen to beak up the land. Urban riders did not realise the importance of a gate left open. In some cases these were considered a right of the old landlords, paying no compensation. A farmer in County Dublin solved the problem on his lands by direct action. He shot the fox as it ran in front of the hounds.

The Law Reform Committee recommended that farmers were liable for damage by farm animals regardless of fault. This recommendation was partially amended.

Pollution

Pollution caused such a variety of problems that the title of the committee was changed from the Industrial and Transport Committee to Industrial and Environmental Committee. On one side compensation was obtained for lead poisoning over 2,000 acres by the Tynagh mines and similarly for poisoning from Silvermines (1989). Bord na Mona caused drainage problems with silting by turf dust. Less definite was pollution from Sellafield and the possible harm to people and stock from high-tension ESB transmissions.

On the other side, polluting farmers were threatened. The risk of prosecution for water contamination was estimated to be high on one third of farms, with an estimated total possible liability for this group of £300 to £600 million. A further 12 per cent of farms had a lower risk, estimated at £100 million liability. The most common danger was from pig slurry and silage effluent affecting drinking water, tourism and fish stocks (22 September 1988). Neighbourhood Watch was promoted as a means of reporting and checking pollution and dumping.

Other Committees

Committees multiplied as the Association grew. By 1993 there were 22 at national level and 15 international. The IFA was represented on 20 outside bodies in Ireland. It has been possible to touch on only a sample of their of their work. Some irritating problems, such as transport restrictions, reorganisation of marketing boards, consolidation of and the dairy industry, ceased to cause concern – whether through EEC influence or IFA representations.

International views or conflicts of interest were thrashed out in COPA. The control of surplus production was a good example. The majority favoured quotas rather than price. The Germans and British preferred price cuts, combined with direct subventions to their own farmers from the national exchequer; the Irish and Danes agreed to quotas, but did not like 1978 as a base year, when Irish output was still growing in response to European prices.

Subsidiary Firms

Some newly felt needs were hived off to independent corporations, usually in co-operative form within the IFA system:

Farm Relief Services were developed in the early 1970s through Macra and in conjunction with ICOS and ICMSA. By 1978 there were 30 associations, employing up to 20 workers each to provide holiday and sickness relief on dairy farms.[20]

FBD provides automatic personal accident cover for all farmers on payment of annual membership fee to IFA. A subsidiary, FBD Life and Pensions, deals with longer term business, life assurance, pensions and investment. There are 50 offices.

Irish Farm Accounts Co-operative gives a specialist farm accountancy and taxation service from 21 offices. The need had been urged by Helge Larsen in lectures to Macra before NFA started. The Danish farm accounts service commenced about 1900. Some Dungarvan farmers adopted the idea. It took the pressure of taxation and the increased technical competence of farmers to spread the service nationwide under the chairmanship of Donal Cashman and the chief executive Peadar Murphy.

These and innumerable other activities realised savings for farmers and increased farm safety to a value far exceeding the cost of subscriptions in any year.

APPENDIX

FINANCE

Financial affairs of the Association evolved with the tasks undertaken.

SOURCES OF FINANCE

Membership fees

Dependence on membership subscriptions was virtually complete till 1970. The simple structure of early years was financed by direct membership fees, based on 2.5 pence per £PLV, with a minimum contribution of 25 pence a farm and a maximum of £10. In 1955 the only income was £12,328, forwarded to head office. County organisations were self-financing. Membership fees rose each year reaching £90,000 in the nine months to September 1965.

Dependence on annual subscriptions from individual farmers is precarious. The individual saw that non-members gained equally with members from price

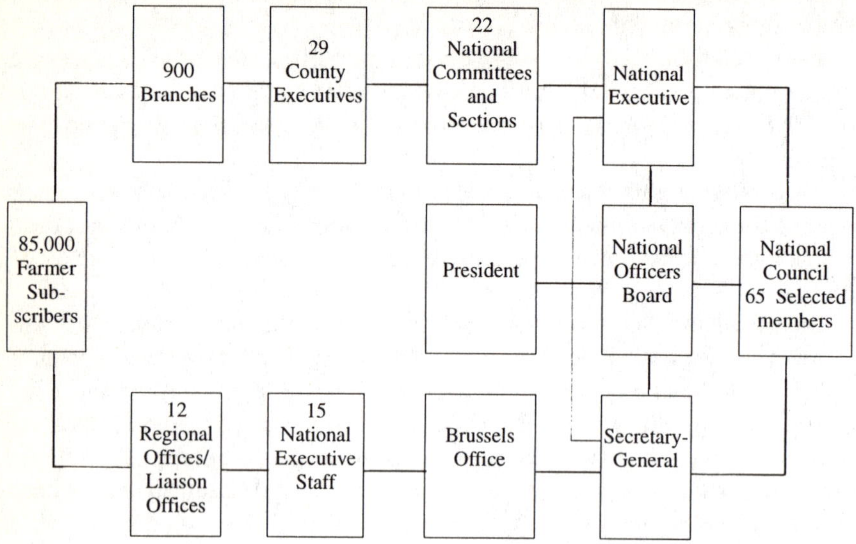

National Committees:

- Livestock Committee
- Dairy Committee
- Grain Committee
- Sheep Committee
- Farm Business Committee
- Sugar Beet & Vegetable Section
- Rural Development Committee
- Liquid Milk Committee
- Pigs & Pigmeat Committee
- Animal Health Committee
- Farm Family Committee
- Industrial & Environment Committee
- Potato Committee
- Animal Feedstuffs Committee
- Salmon Growers' Section
- Western Development Committee
- Poultry Committee
- Horticulture Section
- Horse Committee
- Forestry Committee

The IFA represent your interest on over 20 Outside Bodies including:

Bord Bainne
Bord Glas
CBF
Central Review Committee under the Programme for Competitiveness & Work
Cereals Association of Ireland
COFORD
Council for the Status of Women
County Enterprise Boards
Disadvantaged Areas Appeals Panel
Farm Apprenticeship Board
IFAC
Institute of European Affairs
International Federation of Agri. Producers
Irish Council of the European Movement
Macra na Feirme
National Dairy council
National Economic and Social Council
National Economic and Social Forum (NESF)
National Rural Tourism Co-op
Self Help
Monitoring Committees on the EU Structural Funds
TEAGASC

27 The IFA structure, as described in a 1995 pamphlet

rises or revised regulations. Canvassers, happy to call on their neighbours when fired with enthusiasm for a new organisation or in time of crisis, are reluctant to make an annual 'begging' visit when there is nothing exciting to report. Thrifty farmers wanted to pay a membership fee of £1 by quarterly instalments. Collection by paid organisers was too costly and difficult to supervise. Bankers' orders partly solved the problem for larger farmers with bank accounts but the system was inflexible. With inflation at 20 per cent the contribution had to be raised each year.

The farmers' rights campaign was costly. Funds raised for this purpose were 10 per cent of income in 1966 and 24 per cent in 1967. The most serious cost was in income forgone. The field officers and committee men were so fully engaged in organising of protests and demonstrations that both their farming and their fund raising suffered. A deficit of £300,000 accumulated in the accounts. A special appeal for funds brought balance.

None the less, when Donie Cashman became president the annual deficit was £166,000 (1980). This deficit, 14 per cent of revenue, was of the same proportions as the much criticised government deficit. Such a trend, endangering the life of the association, could not continue.

Development of FBD; raising capital for Cork Marts, Avonmore, creamery reorganisation; campaigning in the referendum for entry to EEC and other valid activities took precedence over promotion of IFA itself. Membership fees stagnated, while inflation increased wage and other costs. Doubling income in 1974 and again by 1976 did not solve the overhanging debt.

Income from other sources was developed till the IFA dependence on membership fees fell to 13 per cent in 1977. The source of income changed again. Membership fees increased steadily in importance from 1985, contributing 46 per cent of revenue in 1992. This increase was due to the Membership Protection Policy introduced with a membership drive in 1986. In the early years members, travelling late on difficult roads, carried their own risk of accident. Several suffered permanent injury. Attempts to insure members while engaged in association business met legal difficulties. Insurers required lists of members from all committees, with dates and times of meetings. A farmer carrying livestock to the mart on his way to a local meeting might not be covered. When a parish hall in Offaly went on fire during a meeting, the member making the booking was held personally liable. His farm would have been sold to meet the costs but for a national subscription. The courts were awarding ever higher damages.

The solution was to cover all members all the time. The Association rules included the whole farm family as members. Therefore, they were all covered for all injuries. While expensive, the cost was £284,000 (6.7 per cent of income) in 1993, an important gap in insurance of farmers handling machinery and cattle was covered. The Association regained its direct relationship with the individual member, endangered by the collection of fees through product levies at creameries and marts. The recurring question of members, 'what benefit do I get above non-members?' was answered. FBD, fulfilling its object of service to members, also gaining entry to the general farm insurance and other business.

Following strong measures of fund raising and expenditure restraint, the accounts have been in surplus every year since 1981. The net liquid position reached £784,000 in 1993. This was 43 per cent of annual membership fees. Fixed assets in buildings and investments were the true wealth of the association.

The 1993 membership fee, including Membership Protection Policy, was £16 plus 35 pence per acre farmed.

Membership

Membership figures of voluntary organisations are always hard to obtain. Who is a member? In the NFA the whole farm family was the unit and had one vote. Any member of the family could represent it at meetings and use that vote. In that sense the number of people members was perhaps three times the number of subscriptions. Does a farmer cease to be a member because his subscription is late or no one asked him for the money in the current year? Does the voluntary contribution paid in levies constitute membership? if so, county figures are misleading because the membership is credited to the county of the mart, not of the farmer. Such figures give no indication of farm size.

This said, there are 900 branches with 57,000 fully paid members. A further 30,000 farmers pay the voluntary levies on commodity sales. There are about 90,000 farms with over 20 Hectares in the Republic.

Other Income

The pattern of financing changed fundamentally in the 1970s. Membership fees fell in importance and the burden of expansion was taken up by contribution from other sources.

Affiliated organisations became important, rising to 26 per cent of income in 1973. This was an indirect membership fee. The Leinster Milk Producers contributed £300 in 1957 and succeeding years. Later contributors were the Beet and Vegetable Growers (1969), the Poultry Producers (1970), the Malting Barley Growers (1971), the Pea, Fruit and Vegetable Growers, Cork Milk Producers, the Irish Commercial Horticultural Association (1971), Irish Wool Producers Ltd (1973), the other regional Milk Producers (1976).

Income from trust organisations developed slowly. The *Irish Farmers' Journal* Trust contribution (£300 in 1956) rose, with fluctuations, to £23,000 (nine per cent of income) in 1973. Insurance commissions were slow to develop. Health insurance yielded a couple of thousand pounds at most. The picture changed with the Farm Business Development Trust, income from which settled at £175,000 a year in 1992 and 1993, four per cent of income.

Minor income arose from the sale of the NFA *Yearbook*. Its circulation of 10,000, contributed a profit as a by-product of NFA publicity service. A National

Discount Scheme started with a flourish. In 1959, 225 firms offered members discounts of between five per cent and 25 per cent. Charges have been made for other services, making little contribution to overheads.

Government payments are important in countries, such as Spain, where vocational organisation is a part of the social and political structure. In Ireland this could not apply. With Sean Lemass (1960) it was agreed that no aid could be given under Irish conditions, even by a willing government. The NFA could not negotiate with its paymaster. The refund of NFA expenditure on foreign travel and organisation were paid in 1963, 1965 and 1966. The highest payment, £3,000 (4 per cent of income) in 1963, covered travel and a contribution to the Consultants' Report under a national efficiency scheme. The last Government aid was £500 in 1966. The Convention in the National Stadium, 1967, resolved that in future no governmental subsidies be accepted.

Funds to buy the head office in Earlsfort Terrace were raised by a lottery. In effect this was an interest free loan repaid to the winners of an annual draw. Sale of this premises and of the lease on the Adelaide Road office contributed £15,000 to building of the Irish Farm Centre.

Product Levies

The introduction of commodity levies fundamentally changed financing in 1974. The IFA became increasingly dependent on produce levies. 49 per cent of revenue came from this source in 1993.

The European Involvement Fund started with the Cork Marts Group in 1972 and extended to other marts next year. Initially introduced to fund the IFA Brussels Office, it was a voluntary levy of 0.1 per cent on sales. (Where the sum collected exceeded £12 the balance was refundable.) Donie Cashman, deputy president, and Dick Hourigan, chief organiser, extended the levy to meat factories and other commodity sectors – liquid milk, grain, pigmeat (1977). Difficulties with the ICMSA delayed collection from creameries till 1980. Levies just balance the Association's expenditure on its foreign activities, as was the intention at their introduction.

Product levies are not a new development in Ireland. The Sugar Manufacturing Act (1933) gave the BGA the right finance its representation of farmers by a levy on beet delivered to the factories. Sean Lemass quoted the BGA example, indicating that he would consider favourably a proposal for such an arrangement from an overall farm representative body. Disunity ruled that out. Membership fees of the LMP had for many years been collected by the milk wholesalers. Since contracts were only given to members of the LMP, this was in effect a compulsory levy. The ICMSA based its finance on fees deducted from the milk cheque. When fixed as a percentage of the milk cheque the finances of the association became secure. The ICMSA bitterly resented the intrusion of the NFA into the creameries and even took legal action against the NFA levy on sales of cattle.

The bulk of funds now come from relatively few sales points rather than individual members subscriptions. Once the farmer agreed to the levy, future deductions were automatic and forwarded in bulk to the association. Payment was clearly linked to the service rendered to the farmer and his volume of production. Income was index linked to farm output and price. It became easier to compile a register of producers to elect members of commodity committees.

At last the members were saved the time and goodwill of house to house collections. ICOS funding was traditionally based on a levy on creamery and store turnover. Few co-operatives paid in full. A large government grant was paid in most years. The organisation would not have survived without it. This gave Government detailed control and made farmers reluctant to pay for a service already paid from tax.

EXPENDITURE PATTERN

The principal cost of a voluntary organisation is the unpaid work of members, particularly its officers. Scattered farms necessitate long journeys to meetings. Work hours are lost and days away from home are the norm. Even out of pocket expenses were often born by the individuals. This is unrecorded and without measure.

In the first year salaries and expenses of staff cost £4,200 from expenditure of £10,700. Macra was paid £1,000 to recoup expenses in launching the NFA. The low expenditure of £1,890 for Council and committees shows the initial dependence on voluntary effort. Expenses fell to £839 in 1956. Meetings were numerous but the the infant organisation lacked the money to pay expenses. The preparation of the Research Institute Report cost £2,200 (18 per cent of income), showing an interesting order of priorities.

Travel

A rural association involves more travel than an urban union, whose members are grouped in towns and gather daily at work. Membership contact becomes more difficult as branches spread. Collection of revenue through marts and creameries cut out the personal contact of canvassers.

Travel expenses of headquarters staff and committees settled to a steady 10 to 14 per cent of revenue. Regional organisation – field staff, county executive and communications – absorbed between 25 and 50 per cent of revenue, in addition to an indeterminate share of headquarters staff time and travel.

An unusual provision is made in the Membership Protection Policy, introduced in 1986 and taking 6.7 per cent of income. The extension of cover to the farm family of paid up members is a sales point in recruiting.

Foreign expenditure appears in the accounts at a modest three per cent in 1958, its share rising little till 1972. Representation in Brussels, travel, permanent office, demonstrations, rose from 2.4 per cent of revenue in 1973 to 20.5 per cent in 1974 and varied around 10 per cent till 1985, currently taking about 5.5 per cent. The levies on produce were introduced to cover IFA expenses in foreign representation.

Sean Lemass, as Taoiseach, accepted the role of the NFA in representing the national interest at foreign meetings of farmers. Had these payments recommenced after the dispute with the Government, the IFA finances would have been sound and farmer contributions greatly reduced.

BALANCING BOOKS

An association must balance its books if it is to survive. As in national government, voluntary organisation policy is shaped by compromise. The spending departments see an unbounded need for beneficial activity, balanced by the finance minister who calls for restraint. Occasionally one or the other becomes dominant, giving rise to neglect of financial balance or to over cautious inactivity. In a vital organisation membership interest brings corrective adjustment of that balance.

In the first year its existence the NFA showed a surplus, relying on the enthusiasm of members and the few staff to provide cheap labour. In spite of increasing income, deficits followed for the next five years, till the deficit for 1960 was almost 20 per cent of income and accumulated deficits 44 per cent. Crisis reform was undertaken. In spite of rising expenditure surplus was shown every year till 1970 (with the exception of 1964). Deficits followed for 8 of the next 11 years till new crisis arose. The NFA could not continue its current operation.

With the reorganisation described above, a secure income of about £4 million a year has led to reserves of some £2 million.

Comparison with the ICMSA

Table 13.1 sets out the main headings of income and expenditure for the IFA and the ICMSA in 1992. It shows that the IFA is the larger enterprise, having seven times the financial resources of the ICMSA. More of its income is derived from direct farm contributions.

When staff and other items are included, just over half IFA expenditure is on international representation – not a heavy item to the ICMSA. Travel expenses within Ireland are heavy, as expected, in the ICMSA as in the NFA.

Both associations showed a prudent surplus in 1992, the ICMSA's being 23 per cent compared with IFA nine per cent.

Table 11 Income and Expenditure Pattern IFA/ICMSA, 1993

	£000	%		£000	%
Income					
Subscriptions	1,773	46		208	38
Product Contributions	2,110	54		339	62
Total	3,883			547	
Expenditure					
Central Office	954	24.6	Salaries etc.	242	44.2
EEC Office & International	238	6.1	Travel	161	29.4
National Council & Committees	370	9.5	Consultancy	14	2.6
Regional Organisation	924	23.8			
Membership Protection Insurance					
Communications & Promotion	305	7.9			
Other Costs	725	18.7		5	0.9
Total	3,516			422	
Surplus	367	9.5		125	22.9

Fitting the New CAP, 1987–1992

THE COMMON AGRICULTURAL POLICY

The gaining of access to foreign markets was a prime object of the NFA from its first year. At first the Common Market Common Agricultural Policy gave all that was promised. The fundamental change was that Irish farmers and the Irish government obtained rights in the making of agricultural policy never previously conceded in international trade. Instead of trying to change a weak situation they struggled to preserve advantages.

Problems of agriculture now centred in Brussels rather than Dublin, but they did not disappear. Technical change and over generous prices generated surpluses which made revisions of policy necessary. New regulations of quality standards, disease, exchange rates made new demands on IFA members and staff. A new expertise had to be developed. Representation abroad came to absorb half the resources of the IFA.

During wartime scarcity all prices were pitched high as an incentive to output but those in Germany far higher than in France or the Netherlands. The political weight of the Bauerverband (German Farmers' Union) made reductions in Germany impossible, so all prices moved up towards German levels. French protests that over-production must result were overruled.

These developments were known in Earlsfort Terrace. The question of quotas or some other control to prevent market gluts was discussed in IFAP in the 1960s. The European Movement held a symposium on the subject, the papers of which were published, in 1976.[1] Membership of COPA from 1972 gave the IFA insider knowledge of developments and plans within the Market. The Mansholt Report, or its interpretation, foreshadowing the elimination of uneconomic holdings, was unpopular. His scheme of slaughter to eliminate the 'postman's cow' (uneconomic production by part-time farmers) was already in operation. Mansholt's actual intention is debatable. When he had finished his report he went to Italy for a wellearned holiday. The publicity section at the Commission in Brussels, wanted good press coverage for the report and highlighted the most spectacular parts. The Mansholt Report actually projected a *reduction* in the rate of decline in farming numbers compared with the years immediately preceding. Mansholt's aim was to control surpluses by substituting a planned redeployment of resources for the unsatisfactory, disorderly flight from the land. Farmers from Land Commission holdings in Kildare and the Meath Gaeltacht greeted him outside the Shelborne Hotel with placards 'Go Home Cromwell' (13 November 1970). When he addressed the Macra Summer Gathering in Killarney Western farmers rolled out a banner 'Go Home Cromwell' and marched out to form a picket. In 1974 the Mansholt Plan was replaced by the a package of EEC socio-economic directives, more suited to development of agricultural regions.

In the medium term the prospect of restraint posed no problem to Ireland. Shortage of fertilisers and low produce prices during the preceding 40 years left farming weak in output and starved of investment. The bewildering increase in milk and beef prices for an unlimited volume gave farmers courage to invest towards a European level. Even if there was no immediate outlet in the European home market the produce could go to Intervention till development was complete. If restrictions on production must come it was all the more important to increase output, establishing a high base from which quotas would be set.

The First Mishap

The Common Agricultural Policy was based on equal prices in all markets, allowing for transport costs. During the period of stable exchange rates before 1973 (under the Smithsonian Agreement) this was achievable.

*The heaviest blow to farming came, unexpectedly, from EEC rules on devaluation, which brought levies on Irish agricultural exports to Britain, the so-called Monetary Compensation Amounts. When Britain devalued by 40 per cent in the 1970s the system could not cope. Britain imported half her food. If the price of imported food increased by the amount of devaluation British workers would make big wage demands to compensate for the increased cost of living. This was avoided by a curious arrangement. Agricultural prices under CAP were only guaranteed in terms of the national currencies. Therefore, British food and agricultural prices remained at the pre-devaluation sterling price. There were now two rates of exchange. The 'Green Pound' remained at the old rate for agriculture. The ordinary ('spot') rate of exchange, which varied from day to day, applied to industrial goods and other transactions.

The British agricultural price was kept below that in other EEC countries by a subsidy from the Commission on imports from the Continent. Similarly, agricultural exports from a devaluing country were taxed to bring them up to the target price in the recipient country. This kept continental farmers happy.

The British did well. Their basic food prices were up to 40 per cent cheaper than in other countries. British farmers had low prices in the market, but that could be compensated by direct subsidies from the British Government. Farmers were only 2.5 per cent of the population, so the tax burden was not intolerable. Food processors, 11 per cent of workers were far more important.

It was a catastrophe for the Irish. On devaluation the foreign exchange value of agricultural exports (40 per cent of net exports) was reduced by the MCA tax. The cost was £100 million a year, £40 a head on cattle in June 1976. (The

Central Bank estimated a cost to Ireland of £3 million a week.) What the farmer bought was governed by import prices, increased by devaluation of the Irish Pound; what the farmer sold was valued at the old, predevaluation, price. Naturally the Irish farmer protested vigorously.

Table 12 Effect of MCAs

Price of Beef, per cwt. live weight, December 1976

Ireland	£26.42
UK	27.05
Denmark	38.05
Germany	46.56
Belgium	47.56

The Irish Government reacted by negotiating a reduction in the value of the Irish Green Pound. The Irish farmer now got a better price than the British, but less than the continentals. The new price difference between the Republic and Northern Ireland (£30 a head on cattle and £6 a pig in October 1976) gave rise to a lively smuggling trade. Some 200,000 head of cattle and 10 per cent of pigs slaughtered in the Republic were smuggled in 1976.[2]

Irish farmers survived – produce prices were still rising to CAP levels. Large sections of the food processing industry died. Due to the MCAs, raw material for food processing (meat, butter, grain, sugar – whether home grown or imported) was cheaper in Britain than in the rest of Europe. Irish processors paid 32 pence a pound for beef but the British paid 19 pence – even for German beef bought at the high price shown in Table 11 above. When processed this food became an *industrial* product, to which MCAs did not apply. The food processing industry of Britain was already well developed and able to exploit the unexpected advantage. Exports to EEC increased from £23 million in 1972 to £314m in 1975 and £1,010 in 1978.[3] On export to Britain Irish beef received an MCA of 4 p. a pound. But when processed as canned beef it became an 'industrial product' and no payment was made.

Alice in Wonderland became the best textbook for explaining the working of the CAP pricing system.

The Irish processors did well at the first stage of production to butter or sides of beef. Creameries thrived and cattle were slaughtered in Ireland rather than exported on the hoof; the number available increased by the smugglers' offerings. These were sold to Intervention at the Green price. The Irish production of more highly processed foods was devastated. MCAs made British raw material about 20 per cent cheaper than in Ireland. Their processors took not only the export but even the Irish home market. Ireland had supplied 10 per cent of the British market in tinned meat. In 1974 the Irish export value in this heading was £9,435,000: in 1977 £115,000. The export market was lost and it became nearly impossible to buy Irish tinned meat in Ireland. Irish cattle or sides of beef were so much dearer than in Britain that processing could not pay. This contributed to the demise of Clover Meats. In the same way sweet manufacturers were all but exterminated. Biscuit exports were hit and the home market only preserved when the Government obtained a derogation, allowing them to compensate for the price difference in flour and sugar by a subsidy from the Irish taxpayer.

Table 13 Impact of MCAs on Beef Processing, 1987

Processed in pence per pound	*Ireland*	*Britain*
Raw material ex Irish factory	80.00	80.00
Processing cost	10.00	10.00
Transport to Britain	3.30	2.50
Net MCA on processed product	- 8.67	0
Net MCA on raw material	0	- 17.35
Variable Premium on raw material	0	- 7.00
Cost of processed product on UK market	84.63	68.15

Source: *Obstacles to Value Added in the Beef Industry*, Irish Meat Exporters' Association, 1987

Ireland had expected to invade the British processed food market when controls and subsidies were removed. Our cost of production was lower and similarities in language, tastes and measures, combined with our ethnic market of emigrants, would ease our entry. Instead the British had these advantages and the bonus that their newspapers and Television advertising penetrated the Irish market without additional cost. Irish industrial wages and taxes grew to be higher than British because Irish food prices were higher. This placed every other industry at a disadvantage

When, in time, the MCAs disappeared the harm had been done. Many firms had been forced out of business. The survivors were so weakened by the unfair competition that they were short of capital. Product research and development could not be undertaken. Fear of another devaluation made new development too risky. It was ten years before a new generation of firms, led by the farmer-controlled, consolidated creameries, could develop.

The IFA had concentrated on the farm gate price. They strongly pressed for a full devaluation of the Green Pound. Irish Ministers sought and obtained successive devaluations of the 'Green Pound' to reduce the MCA tax on exports. By increasing Irish farm produce price above that in Britain they almost destroyed the Irish food processing industry with its employment, its value added and more secure markets. Employment in Ireland was reduced, many foods require more labour in processing than in farming.

On the other hand farmers could hardly accept a further 20 per cent reduction in price when costs of farm inputs and wages had been allowed to rise. It might have been wiser to allow the Irish pound to rise in value against sterling. In the long term, and with hindsight, the decision to favour the farmers by negotiating a lower Green rate of exchange was unwise. True, it gave a better farm gate price. Cattle were slaughtered in Ireland rather than exported as stores. Slaughter-houses were built and the owners prospered. Markets were no longer so dependent on demand for store cattle in Britain, where the weather of 1974 caused a fall of 24 per cent in Irish cattle prices. However, the reliance on bulk sales to Intervention inhibited the development of further processing and of market contacts. The value added was disappointing. In the reorganisation of CAP and GATT the trade was vulnerable.

As Table 14 shows, employment in the food and drink manufacture was 28.6 per cent of all manufacturing in Ireland. NFA forecasts had shown an increase of 20,000 jobs in this sector.[4] In fact there was a reduction of 10,000.

Table 14 Employment in Food Processing as a Proportion of All Manufacturing Employment, 1986

Country	*% All Food*	*Meat, Milk, Milling*
Ireland	28.6	13.2
Denmark	20.1	9.6
Netherlands	16.6	5.4
UK	12.1	3.3
Belgium	11.2	2.8
France	9.7	4.5
Italy	8.1	2.6
Germany	6.6	1.5

Source: Eurostat, *Structures and Activity of Industry*, 1986/7

It does not follow from this disappointment that Ireland's entry to EEC was a mistake. The alternative 'World Prices' obtained by New Zealand and other outsiders were a fraction of those in Europe's sheltered market. Free trade in manufactured goods brought industrial expansion impossible for a small isolated state. There was a clear net gain.

IFA Policy on CAP

Government and farmers were broadly agreed on foreign policy. The Irish took no initiative for reform of CAP because they feared that any change would be for the worse. In discussion with the Department of Agriculture the NFA recommended that the increase in prices to continental levels should be spread over three years longer. The shock to the markets would be less. The political target was the immediate price increase and the postponement of quotas for as long as possible to allow increased output to be established. Efforts by farmers and Government to abolish MCAs were unsuccessful. The IFA National Council discussed breaking the Sterling link a number of times but, as Lorcan Blake, NFA economist, pointed out (June 1976) this could not work when Irish inflation was even higher than the British.

CAP Reform Phase 1, 1980–6

IFA president, Paddy Lane, in his AGM address 1977, saw that 'If there is a problem of over production in dairy produce in the Community, measures will have to be taken, but not through reducing prices.' He proposed a non-marketing premium to reduce production and encourage farmers to switch from milk to beef and grain.

28 Mark Clinton's negotiation was admired (an *IFJ* cartoon).

The Commission redirected production by a 'co-responsibility levy' on milk, doubling the premium for suckler cows and a common policy for sheep meat.

When EC expenditure reached its budget ceiling in 1983 policy had to be reviewed. The main Commission proposal was a quota system, to be implemented by a super-levy. Milk production was to be cut back to that of 1981 plus one per cent. The IFA opposed these measures of two main grounds: 1) The Irish dairy expansion was adjusting rapidly to reflect its natural advantage in the competitive conditions of the Common Market. Milk yield, for example, had increased to 700 gallons a cow, catching up on the EC average 880 gallons and expected to reach the 1,000 gallons of Netherlands, Germany, UK and Denmark. To cut the quota to 1 per cent above 1981 would mean a reduction of 20 per cent from the expected output of 1984. 2) Ireland had a 'vital national interest' because milk production was six and a half times as important to the Irish economy as in the overall EC economy.

The IFA proposed a packet of measures: 1) abolition of MCAs. 2) Control of cereal substitute imports, which were glutting the market. 3) a tax on vegetable oils and fats. 4) limitation of New Zealand butter imports. 5) a levy on intensive milk production within the Community. Alan Gillis, future IFA president, raised a lone voice at the National Council meeting, August 1980, asking that 'the pros and cons of quota systems should be examined'.

At this stage the IFA campaigned against the quota system. Members were informed of the IFA policy and stirred to mass meetings, at which ICMSA and trade unions participated (Thurles, December 1983). The Taoiseach, Dr Garret FitzGerald, gave the farmers' case high priority at EC policy meetings and by direct representations to President Mitterand, of France and Dr Kohl, Chancellor of West Germany. The IFA made also its presence felt at EC decision making meetings in Athens and elsewhere.

The Council of Ministers decided, in March 1984, to hold their decision on quotas for most countries at one per cent over 1981, but to give Ireland a quota of 4.6 per cent over 1983 to accommodate the expected expansion. This quota was about 20 per cent higher than under the 1981 formula of one per cent.

CAP Reform, Phase 2, 1986–1988

The milk quota decisions of 1984 did not solve the over-production problem. Land taken from milk production was transferred to production of other products. A corset on one part of the body causes unsightly bulges elsewhere. TJ Maher had pointed out years earlier that the storage cost of surpluses absorbed much of the money intended for farm income support. Switzerland's direct payment system attracted favourable attention in Germany. As principle paymasters the Germans saw that price support for unwanted produce was inefficient in raising farm incomes. Storage was costly and subsidised exports both expensive and harmful to foreign farmers.

The Irish viewed any scheme of direct payment to farmers with alarm. In the days of Margaret Thatcher it seemed impossible that Britain would agree to a cash transfer to farmers in other countries. There would no longer be a common market if richer national governments gave direct payments which the exporting countries could not match.

In July 1985 the Commission published a 'green paper', discussion document 'Perspectives for the Common Agricultural Policy'. This was followed by proposal for a payment to farmers ceasing milk production; co-responsibility levies (three per cent) for cereals and reduced Intervention support for both cereals and beef.

The Irish had a special problem with quotas. Dairy farms were responding to the recent price increases which brought them to equality with other EC countries. Quotas, as at first proposed, would tie them back to the old output. The IFA canvassed the case for a larger milk quota in IFAP and COPA. Other farm unions, in particular the German Bauernverband (DBV), agreed that Ireland, having the best conditions in Europe for dairying, should get a special quota to allow adjustment. Their support with their governments was generous and effective in obtaining a 10 per cent quota increase.

The IFA reversed its policy at the 1986 AGM. Joe Rea, president, stated the new approach: 'Surely the solution is the application of supply management to agricultural production. The EC now has over 300 million consumers, surely it should be possible to match food production to EC requirements.' The IFA realised that supply had to be aligned to demand. Farm income would be better protected if supply was controled by quota than if it was forced down by price cuts.

Control of output by quota on output and withdrawal of land from food production by 'set aside' of land, afforestation and other of inputs quite changed the agricultural scene. The old emphasis on increasing production, the pride of the old farmers, was washed away by surpluses. The individual farmer depended on the power of his pressure group more than on the market. In 'social agriculture' negotiation of favourable terms became the first requirement. It was a precarious dependency. As Alan Gillis put it, 'What happens when they don't like the look on your face?'

GATT

Forty years back the Irish farmers complained that they, with the Australians and New Zealanders, were excluded from high priced industrial markets. With membership of the European Union the Irish view changed. The managed market they now favoured came under threat from the extension of the General Agreement on Tariffs and Trade to agricultural products in the Uruguay Round of negotiations, started in 1987. The Blair House Agreement (1992) signed by Commissioner Ray MacSharry and Edward Maddigan, US Secretary for Agriculture, and would result in more extensive cuts in agriculture than those in the CAP. A major anti-GATT rally was held in the National Stadium (December 1992) addressed by Mr Kjeldsen, Danish president of COPA, and the president of the French union FNSEA. All living past presidents of the IFA were present, except Paddy Lane, who declined the invitation because of his association with Ray MacSharry. The other farm organisations, Macra, ICOS and ICMSA, were also represented. IFA president Alan Gillis, attended the World Family Farmers' Summit meetings in Tokyo, which unanimously rejected the Blair House

Agreement. Alan Gillis, as vice-president of COPA, met the Japanese Prime Minister, who was chairman of the 'G7' summit of industrial nations, to argue for renegotiations in GATT to safeguard farm families and check rural depopulation. The final Uruguay Agreement in December 1993 confirmed that of Blair House.

These last developments go beyond the scope of history. The outcome is unknown. What is clear is the progressive development in farm organisation from the individual owner to local, and national regulation, then to European Continental and further to world-wide controls. Opinions must vary about the desirability of current changes in technology and trade. None can contest the need for organised farmer representation in decision making and in assisting the individual farmer to understand and adapt to rapidly changing circumstances.

Review

A century is a mere blink of history. In that time Ireland has experienced a series of revolutions, in politics, technology, economics, education.

An underdeveloped social organisation made adaptation more difficult. In 1850, after the Famine, society had virtually no structure. The population, reduced by 22 per cent in ten years, was demoralised. Community effort was first directed to the most pressing needs of reconstruction, religious emancipation, education, land reform, trade unions. National independence was achieved in 1922 without a social philosophy. The network of vocational and other organisations, each with its recognised rights and duties, which shape the internal structure of the state in Continental countries, was missing. There was no accepted aristocracy to give leadership.

The newly formed state naturally expanded to fill the vacuum left by the retiring British government. State and Church, the only strong authorities, became dominant. The Church shed its non-essential work, such as leadership in co-operatives, as soon as it could. It was more difficult to find a boundary to the functions of the State. Individual farms were too small and scattered, the education of the owners too low, to generate spontaneous development in technology or social structure. In consequence, the State took an exceptional degree of control, not only over prices but in the smallest details of production, breeds of stock, technical education.

Contrary to the stated intention of its founders, the Department of Agriculture by-passed the existing farm organisation. The political power enjoyed by the Ministers, and through them by the civil servants, was enormous. Neither was willing to abdicate any part of its territory to subsidiary organisations. Politicians viewed each development as a party political move; civil servants considered the State the natural leader of development. Control of social functions – Education, Health, Local Government, Agriculture – were progressively centralised.

After World War 2 a crisis came. New faults in the system became apparent in production, marketing, quality control, processing. Policies in transport, industrialisation, taxation were ill adapted to efficient farming. Farm incomes fell, productive investment in agriculture and in industry was very low, emigration concealed unemployment.

By the 1950s it was apparent that the system had not worked. Irish farming had fallen so far behind the leading countries that fundamental change had to come by forced marches for survival.

Even if the state-centred system had been efficient the new generation would have been restive at the prevailing paternalism. In the circumstances friction was inevitable. Ministers appointed through the party political system were not usually in touch with technically advanced agriculture. The administrative civil servants dominated the technical officers and were conservative by their selection and the nature of their work. No guiding aim was apparent.

Rural society has adapted to the new technology of production, transport, marketing. The rising generation in rural areas were better educated than their parents. Even more important were the social and educational clubs which spread throughout the country. These performed for Ireland the work of the folkhighschools of Denmark. An articulate and informed citizenry; with knowledge of society in other countries; having close contact between their groups, grew to maturity.

The organisational structure proposed by Macra na Feirme in 1954 rested on three associations:

- an educational and cultural group, Macra na Feirme
- a business grouping of co-operatives, the Irish Co-operative Organisation Society
- a professional group, the National Farmers' Association.

Relations with the Government were close at the start and under Sean Lemass as Taoiseach. He planned to recognise a single representative organisation for each social group with which he could formulate policy.

Farmers' frustration with declining incomes and Government inaction in many areas led to a major confrontation from 1965 till 1970.

When relations with Government resumed, the atmosphere was changed. Farmers and Government had a common cause to argue within the European Community. Up to 1978 produce prices, rising to Continental levels, removed tensions. Adjustments in marketing gave farmers most of what they had previously been demanding.

To avail of the new opportunities Ireland's internal organisation changed. The co-operatives combined into a half dozen major international companies (PLCs) with business turnover of about £1 billion each. A COGECA survey found that Ireland was the only EC member country giving co-operatives no tax or other financial advantage in 1992. The NFA became the IFA, combining nearly all the farm organisations – with the unfortunate exception of the ICMSA. Government and farmers had a common international purpose. The IFA has as many committees dealing with Brussels as with purely national affairs. Growth of staffing and expenses marked its growth to maturity as an established part of society in Ireland and abroad.

The new generation of farmers are far removed from their fathers. They are more highly specialised and technically competent. Knowledge of regulations and grant application forms have become vital for survival. Nonetheless, the Teagasc National Farm Survey 1994 found only 44% of farms viable – yielding agricultural wage plus 5% return on capital other than land. The rural problem is not yet solved.

This history maps the change in a way of life.

> 'Do scríobhas go mion-cruinn ar a lán d'ár gcúrsaí d'fhonn go mbeadh cuimhne i mball éigin ortha agus thugas iarracht ar mheon na ndaoine bhí im' thimcheall a chur síos chun go mbeadh ár dtuairisc 'ár ndhiadh, mar ni beidh ár leithéidí arís ann.'
>
> Tomás Ó Criomhthinn, *An t-Oileánach*

Chronology

1889	First co-op creamery.
1894	IAOS founded.
1900	Department of Agriculture.
1916	Irish Farmers' Union.
1926	Beet Growers' Association.
1931	United Farmers' Protection Association.
	Land Annuities Association.
	Muintir na Tire.
1935	Liquid Milk Producers' Associations.
1937	Irish Farmers' Federation. Milk strikes.
1944	Macra na Feirme.
1952	Irish Creamery Milk Suppliers' Association. Milk strikes.
1955	NFA foundation meeting in Four Provinces, Dublin.
	450 branches formed; membership 12,000 (March).
	First NFA National Council meets (June).
1956	Join International Federation of Agricultural Producers.
	First auction marts.
	NFA/Macra Report on Foras Taluntais
1958	Greene Marketing Committee Reports.
1958	First Annual Review.
1960	IFAP World Conference, Bray.
	National Liquid Milk Producers' Association formed
1962	15 farm organisations in NFA
	First National Convention.
	Protest marches in Letterkenny, Waterford etc.
	Irish Farmer and the Common Market published.
1963	Introduction of National Farmers' Insurance.
1964	Launch of Farm Apprenticeship Scheme.
1965	Rickard Deasy, president, NFA.
1966	Farmers' Rights Campaign. March to Dublin.
1967	Machinery demonstrations; commodity strike;
	100 farmers jailed; seizure of stock; Commodity Strike.
	Government's National Agricultural Council.

1968	Marts Act.
	2,500 farmers attend National Convention in Stadium.
	Launch of FBD.
	Formation of Agricultural Trust.
	For Oige.
1969	BGA joins NFA.
	TJ Maher President, NFA.
1970	Publication of Green Book.
1971	LMP joins NFA.
1972	Formation of IFA.
	Opening of Brussels Office. Observer status in COPA.
	Opening of Irish Farm Centre.
1973	Membership of EEC and of COPA. IAOS in COGECA. Takeover of Dairy Disposal Co. by Kerry Co-op.
1976	Patrick Lane, president, IFA.
1977	Agricultural Relations Bureau formed.
	Nenagh Agreement. IFA/ICMSA Working Group.
1979	IAOS becomes ICOS.
1980	National Co-operative Farm Relief Services Ltd.
	Donal Cashman, president, IFA.
	Irish Farm accounts Co-operative.
	NESC membership.
1981	Way-leaves obtained from ESB, Posts & Telegraphs, waterworks, Bord Gáis.
1982	Abolition of Rates
1984	Joseph Rea, president, IFA.
1986	Kerry Co-op forms first PLC.
1988	Tom Clinton, president, IFA.
	Agriline market reports.
1990	Alan Gillis, president, IFA.
1993	John Donnelly, president, IFA.

Notes

INTRODUCTION

1 'Three Years of Hard Labour', Address to the Irish Society of Oxford, 31 October 1924, quoted in Boyce, D.G., *Nationalism in Ireland* (London, 1995).

CHAPTER 1

1 *Ireland in the New Century*, 1904, reprinted Dublin, ed., 1982. p. 44.
2 Details given in George Russell (AE), 'A Plea for Justice' and 'Thoughts for British Co-operators', *Irish Homestead*, 1921.
3 J. Johnston, *Post-War Planning in Irish Agriculture*, Irish Independent Pamphlet, 1943.
4 Annual Report of the IFF, *Farming in Ireland*, 1948/9.
5 A detailed history of these developments is given in *Milk to Market* by Doyle and Smith, Leinster Milk Producers' Association, 1989.
6 Address to Macra, Robinstown, 1952.
7 Whyte, J.H., *Dáil Deputies*, Dublin, 1966.
8 Address to Macra, quoted in *IFJ*, 16 October 1954.
9 *Report of the Recess Committee on the Establishment of a Department of Agriculture and Industries for Ireland*, Dublin, 1896.
10 Hansard, 8 June 1896.
11 Section 11.
12 Section 5.
13 Section 13.
14 Daniel Hoctor, *The Department's Story*, Dublin, 1972, p. 56.
15 Anderson, R.A. *With Horace Plunkett in Ireland*, reprinted, Dublin, 1983.
16 For details see Bolger, P., *The Irish Co-operative Movement*, Dublin, 1977 or Anderson, R.A., op. cit.
17 Johnston, J., *Agricultural Co-operation in Northern Ireland*,1969.
18 The system in each EC country is described in R-Europa, *Les Syndicats Agricoles en Europe*, Paris, 1992.
19 *Journeys to England and Ireland*, ed. Meyer, London, 1958.
20 *IFJ*, 11 August 1976.
21 *Report of the Commission on Vocational Organisation*, 1943, par. 222–3.
22 Ibid., par. 225.
23 IBEC Technical Services Corp., *Appraisal of Ireland's Industrial Potential*, New York, 1952.
24 Speech to Wicklow Comhairle Ceanntair, Rathdrum.
25 DD, 5 April 1953.
26 *Irish Times* report, 8 March 1956.

CHAPTER 2

1 Murphy, M., 'Financial Results on Mixed Dairy Farms in 1942/3 as Compared with 1938/9', *JSSISI* vol. XVII. p. 269ff.
2 Smith, L., 'Recent Developments in Northern Irish Agriculture', *JSSISI*, 4 March 1949.
3 Holmes, G.A, *Report on the Present State and Methods for Improvement of Irish Land*, Dublin, 1949.
4 At p. 59.
5 *Reports on Agricultural Policy*, Dublin, 1945 (P. 7175), sections 89 & 207.
6 Smith, L.P. F., 'Productivity in British Agriculture', *Manchester School*, January 1950.
7 OEEC, *8th Report*, vol. 2 ch. 8.
8 The Coal-Cattle Agreement of 1935; the trade agreements of 1938, 1948 & 1960

culminating in the Anglo- Irish Free Trade Area Agreement, 1965.
9 PEP, London, *Report No. 14*.
10 O'Brien, J. (ed.), *The Vanishing Irish*, London, 1954.

CHAPTER 3

1 Letter to Lord Selkirk, in *The Black Book of Edgeworthstown*.
2 See Dowling, P.J., *The Hedge Schools of Ireland*, Dublin, undated.
3 Letter of John O'Dwyer, *IFJ*, 25 August 1955.
4 O'Connor, R., & Keogh, P., *Crisis in the Cattle Industry*, ESRI Broadsheet No. 13, 1975

CHAPTER 4

1 NC, 8 November 1976.

CHAPTER 5

1 Cunningham, E.P., 'Genetic Improvement of the Irish Cattle Population', *JSSISI*, 1965–6, p.101. The same issue contains comments on the difficulty of dual purpose breeding by Prof. John MacCarthy on pp. 127 et seq.
2 See Professor J.O'Donovan. *Economic History of Livestock in Ireland* (Cork, 1940), 373.
3 Warbles: T.O'Nuallain; Brucellosis: T. Twomey and P.P. Lee; Scab: J. Hannan.
4 Health and Disease Sub-Committee Report, 1963.
5 *Farmer's Gazette*, 13 September 1958 & 4 October 1958.
6 *IT*, 2 September 1958
7 O'Connor & Keogh, op. cit. p. 4.
8 Cf. Smith, L., *Agricultural Ireland*, February 1951.
9 *Kerryman*, 21 February 1959.
10 O'Gorman, *Sunday Press*, 8 October 1959.
11 NE, 18 December 1957.
12 Ballymagelligot Case, 1919, AC, 5483.
13 See Bolger, Patrick, *The Irish Co-operative Movement* (Dublin 1977), pp. 203–4 & 219–20.
14 Source: NC.
15 NE, 10 October 1960.
16 Senior, B., 'Pig Production Methods, Denmark, Sweden Holland Ireland', *JSSIS*, 1949.
17 M. Delehanty, Chairman's Report, *NFA Year Book*, 1962.
18 Cf. Hoctor, p. 257.
19 Meeting with NFA, 16 April 1956.
20 Press release, 15 August 1956.
21 *Advisory Committee Report*, para. 57 on.
22 Johnny Callinan, Tom Beegin, Antony O'Dea, Johnny Casey, Dick Langford, Sean Healy.
23 *IFJ*, 21 January 1956.
24 NE, 12 October 1961.

CHAPTER 6

1 31 December 1955.
2 *Irish Independent*, 17 September 1954.
3 AGM, May 1958.
4 *Fact Sheets on the European Parliament and the Activities of the European Union* (European Parliament, Luxemburg 1993), pp. 294–7.
5 Memo to ICMSA, 16 January 1963, in the National Library.
6 Blackwell, J., Convery F. et al., *Natural Resource Application and State Enterprise.* Resource and Environment Policy Centre, UCD, Policy Series No. 1, 1983.
7 DD, 1 March 1956.
8 *Committee on Industrial Organisation* (Dublin, 1963), pn. 6927.
9 T. Finegan, R. MacCulloch & J. Monahan.
10 NFA press release, 22 February 1958.
11 NFA, 1986.
12 *Report of the Commissioners Appointed to Inquire into the State of the Fairs and Markets*, 1855, Cmd 1674; *Report of the Royal Commission on Marketing Rights and Tolls*, 1891. Cmd 6265, xxxvii.
13 *Sunday Press*, 19 December 1993.
14 W. Norton, SD, 18 April 1956.
15 Louis Smith was present.
16 Memo of IAOS, 1953.
17 Report, 1957.
18 Letter of P.G. Byrne, Commercial Manager, 13 June 1962.
19 SD, 18 April 1956.
20 NC, 15 May 1969; 21 November 1973.
21 IT, 2 January 1957.
22 Isles, L., & Cuthbert, N., *An Economic Survey of Northern Ireland*, (Belfast, 1957) chap. X.

23 Branigan, D., 'Survey of Ireland's Shipping Industry', talk to National Agricultural & Industrial Development Association, January 1957, Mimeo.
24 Commission on Taxation, *First Report*, 1982.
25 Walker, Prof. D., *Local Government and County Incomes*, ESRI Paper 18, March 1964.
26 12 April 1960 and following issues & *IFJ*, February 1964.
27 Stationery Office, 1945.
28 'Green Book', p. 144.
29 Gilmore, F., *A Survey of Agricultural Credit in Ireland* (Dublin 1959).
30 Compiled by Brendan Power; published by NFA,1959.
31 Smith, L.P.F., *Comparison of Certain Finance Costs in Agriculture,* UCD, 1965.
32 Agricultural Commission, 1923; Banking Commission, 1926, Second Interim Report.
33 Gilmore, *Survey*, op. cit., p. 8.
34 Gilmore, introducing his report at UCD.
35 Smith, L., op. cit., Tables 6 & 7.
36 NFA, January 1964.
37 Sheehy, S., & O'Connor, R., *Economics of Irish Agriculture* (Dublin, 1985) p. 39.
38 NC, 27 March 1957.
39 NE, 16 May 1961.
40 Smith, L.P.F., *Comparison of Certain Financial Costs in Agriculture*, Chapter IV, UCD, 1965.
41 *Commission on Emigration and Population Problems* (Dublin, 1955). s. 182.
42 3rd Report, section 51.
43 Section 44c.
44 Senator Maurice Manning, script of lecture, 12 February 1956.

CHAPTER 8

1 *Year Book*, President's Message, 1960.
2 NC, 20 July 1960.
3 Smith, L., *The Common Market and the Irish Farmer*, NFA Publication No. 21, October 1960.
4 *The Echo*, New Ross, 2 June 1962.
5 *Irish Agriculture in the EEC*, ICMSA Mimeo, 114pp, 1961. National Library.
6 *Nongovernmental Organisations for European Co-operation* (Documentation Centre Strasbourg, January 1954).
7 Smith, L., Shelbourne Hotel, 1 September 1960.
8 NC, 8 March 1972.
9 *The Irish Farmer and the European Community*, Irish Council of the European Movement, Occasional Paper, 1972.
10 Section 7.
11 To Cork Co. Executive, April 1960.

CHAPTER 9

1 *Pope John and the Farmer*, NFA, 1963.
2 Quoted Smith, C.F., *History of the Royal Irish Automobile Club* (Dublin, 1994), p. 165.
3 NE, 24 October 1964.
4 NE, 19 June 1967.
5 Foras Taluntais, Dublin, 1969.
6 *IFJ*, 2 July 1977.
7 Galbraith, J.K., *American Capitalism: The Concept of Countervailing Power* (Boston, 1952).

CHAPTER 9

1 Chubb, B., *The Government of Ireland* (Dublin, 1974 ed.), p. 32.
2 Coakley, J. & Gallagher M., *Politics in the Republic of Ireland* (Galway, 1992).
3 Ibid., p. 100.
4 Ibid., p. 170.
5 Speech to Macra gathering, Robinstown, 13 October 1954.
6 DD, 4 April 1966.
7 NFA press release.
8 Dr Henry Kennedy, Paddy O'Keeffe, Jim Byrne, Louis Smith.
9 *IT & Cork Examiner*, 11 January 1962.
10 Killarney speech.
11 A. de Tocqueville, *Ancien Régime*.
12 Letters on file.
13 *Irish Press*, 1 February 1932, quoted in Doone, 1988.
14 *Quadragesimo Anno*, 1934.
15 Engineers' Hall, 5 October 1956.
16 Lee, J., *Ireland 1912–1985* (Cambridge, 1989), p. 390.
17 Quoted MacCarthy, C., *Trade Unions in Ireland* (Dublin, 1977), pp. 265–6.
18 Speech to Comhchomhairle, Átha Cliath Fianna Fáil, Shelbourne Hotel,

24 July 1964, on the Second Programme for Economic Expansion.
19 Encyclical, *Mater et Magistra*, Para. 137 & 138.
20 Scully, J., *Irish Times*, 3 November 1954.
21 1958. See chapter 5.
22 Lee, J., op. cit., p. 271.
23 Present were the Minister, Patrick Smith, Dr Greene, Messers J. Callinan, J. Buttimer, T. Finegan, M. Gibbons, J. Mooney with S. Healy and Dr L. Smith.
24 Dated 29 January 1964; see Appendix to chapter 9.
25 NE, 14 February 1965.
26 Press release, NFA, Mullingar, 14 November 1963.
27 *Nationalist*, 25 November 1961.
28 *Wicklow People*, 2 December 1961.
29 *Cork Examiner*, 19 February 1962.
30 Autumn 1961, Tramore, quoted in *Sunday Independent*, 30 July 1967.
31 *IFJ*, 30 September 1968.
32 The Agricultural Association of Ireland reserved the function of wage negotiation when it affiliated to the NFA.
33 O'Connor & Keogh, op. cit.
34 Pr 7697.
35 Ibid., p. 196.
36 Haughey was a member of the Horticultural Association; all the members of which had become automatic members of the NFA; Dillon and Smith were honorary members of the NFA.
37 Opinion, 27 February 1964.
38 Opinion, 7 April 1967.
39 NE, 25 March 1966.
40 NC, 26 August 1965.
41 NC, 23 September 1965.
42 26 March 1966.

CHAPTER 10

1 NC, 15 September 1966.
2 NE, 21 April 1966.
3 A detailed study of relations between Government and farm unions in the countries of the European Union is contained in *Les Syndicates Agricoles*, op. cit.
4 *IFJ*, Editorial, 24 September 1966.
5 *IFJ*, 4 March 1967.
6 *IT*, 18 October 1966.
7 DD, 26 September 1966.
8 DD, 3 November 1966.
9 DD, 5 February 1970.
10 NE, 9 November 1966.
11 DD, 3 November 1966.
12 *IFJ*, 7 January 1967.
13 *IFJ*, 14 January 1967.
14 Charles Haughey, DD, 3 November 1966.
15 Inquiry through clerical channels found no records of the whole incident.
16 Letter, 11 March 1967.
17 FitzGerald, G., *Planning in Ireland*, (1968), p. 227.
18 *IFJ*, 11 February 1967.
19 *IFJ*, 11 February 1967.
20 *IT*, 31 May 1967.
21 Lynch statement to Sean Healy.
22 NC, 12 March 1968.
23 DD, 13 November 1968.
24 Cooke, H., and Sprague, G., *Irish Dairy Organisation* (Dublin, 1968), p. 131.
25 NC 19 September 1967.
26 Conversation in the Gents toilet of Gresham Hotel during the election.
27 Press release, 12 September.
28 NC, 24 September 1968.
29 NE, 22 January 1969.

CHAPTER 11

1 DD, 22 March 1972.
2 AGM, 1971.
3 O'Connor and Keogh, op. cit., p. 30.
4 A more detailed analyses can be found in *Economics of Irish Agriculture* by Sheehy and O'Connor – on which neutral source much of the above is based.

CHAPTER 12

1 NC, 19 September 1967.
2 Minutes of National Liquid Milk Group.
3 Professor Lyons, *Agricultural Ireland*, 1954.
4 Smith, L. , *Agricultural Ireland* and elsewhere in 1950–2.
5 See the case of Town of Monaghan Creamery in chapter 1 above.
6 See contribution of J.T. O'Brien, General Secretary UFU, to *NFA Year Book*, 1962.
7 County Committee for Agriculture, Dungarvan, 18 March 1950.

9 Reproduced in *ICMSA 1950 – Celebrating 50 Years Serving the Family Farmer*, 1990, p. 106.
10 Letter of John O'Dwyer, *IFJ*, 25 August 1955.
11 Rule IX.7.
12 Rule IX.10.
13 *IFJ*, 2 April 1955.
14 *Nationalist*, 26 March 1954.
15 IP, 20 January 1956.
16 *Limerick Chronicle*, 2 April 1959.
17 *Kerryman*, 11 April 1959.
18 NE, February 1967; December 1972; October 1967; May 1975 etc.
19 Hotel Ormond, Nenagh, 24 March 1964.
20 *Irish Farmer*, April 1967.
21 IP, 30 October 1968.
22 NC, August 1974.
23 Inaugural speech, 19 March 1976.
24 NC, 1 June 1976 & 5 October 1976.
25 NC, 14 April 1975.
26 *IFJ*, 4 March 1978.
27 Leddy, *IFJ*, 18 November 1978.
28 Reports in *IFJ*, 2 August 1978 and *Irish Independent*.
29 *Report* of Two-Tier Milk Price Study Group, 1967, Dublin, Pr. 9673.
30 *ICMSA Forty Years celebration publication*, 1990, p. 21.
31 Limerick meeting, March 1972.
37 M. Chirac, like some others in positions of influence, was friendly to Ireland from student days.
38 Stated in 'ICMSA Official News and Views', *Farmers' Gazette*, 25 June 1960.

CHAPTER 13

1 Inaugural speech, 19 March 1976.
2 *IFJ*, 5 February 1977.
3 Hussey, Gemma, *Ireland To-Day* (Dublin, 1993), p. 361.
4 Arnold, B., *Haughey* (London 1993), p. 146.
5 *First Report of Commission on Taxation*, 1982, Table 26.3, p. 590.
6 NC, 5 December 1978.
7 *IFJ*, 30 September 1978.
8 Dr Crotty, consultant economist to ICMSA, advocated a high land tax as a matter of principle. In 1966 he wrote: 'A land tax would stimulate agricultural production in other ways, apart from improving the allocation of land among farmers ... the farmer would pay a heavy land-tax, comparable with con-acre rents or the rents paid to landlords in the nineteenth century': *Irish Agricultural Production* (Cork, 1966). In Chapter X he argued that such a tax would take land from inefficient farmers and reallocate it to the efficient, so increasing output. This was not ICMSA policy, but has some similarity.
9 Committee minutes, 9 August 1985.
10 Sanford, C.T. et al., *Costs and Benefits of VAT* (London, 1980).
11 See also the special tax on co-operatives in Transport section.
12 14:7:14; 10:10:20 and 18:6:12.
13 AGM Report 1982 & *IFJ*, 10 March 1984.
14 AGM Report 1981.
15 *IFJ*, 7 October 1978.
16 *IFJ*, 18 November 1978.
17 *IFJ*, 16 December 1978.
18 *IFJ*, 16 September 1978.
19 *IFJ*, 22 September 1978.
20 *IFJ*, 30 September 1978.

CHAPTER 14

1 Irish Council of the European Movement, Dublin; also, from the European League for Economic Co-operation: Binswanger and others, *Die Europaische Agrarpolitik vor Neuen Alternativen*, Haupt 1977.
2 See Norton, D.A.G., *Ireland, the Cap, Trade Distortion and Induced Smuggling Activity* (Dublin, 1983).
3 Lomax D., 'UK Food Industry and the EEC, *Westminister Bank Review*, 1978.
4 *Irish Farming and the Common Market* (NFA and Irish Council of the European Movement. 1972), p. 9.

Index